T0201378

MATHEMATICAL MODELING
IN SCIENCE AND ENGINEERING

MATHEMATICAL MODELING IN SCIENCE AND ENGINEERING

An Axiomatic Approach

ISMAEL HERRERA

GEORGE F. PINDER

A JOHN WILEY & SONS, INC., PUBLICATION

Library of Congress Cataloging-in-Publication Data:

Herrera, Ismael.
    Mathematical modeling in science and engineering : an axiomatic approach / Ismael Herrera, George F. Pinder. — 1st ed.
        p. cm.
    Includes index.
    Summary: "This book uses a novel and powerful approach to teaching MCM, the Axiomatic Approach, which permits incorporating in a single model, systems that occur in many different branches of science and engineering" — provided by publisher.
    ISBN 978-1-118-08757-2 (hardback)
    1. System analysis — Mathematical models. 2. Science—Mathematical models. 3. Engineering—Mathematical models. I. Pinder, George Francis, 1942– II. Title.
    QA402.H45 2012
    501'.51—dc23   2011036329

Printed in the United States of America.

10 9 8 7 6 5 4 3 2 1

Dedicated to:

The National University of Mexico (UNAM) where I learned the paradigms of mathematical thinking: generality, clarity and simplicity;

and to Brown University where I learned to use them in the solution of problems of practical interest.

IHR

My esteemed students, mentors and colleagues who have stimulated and nurtured my interest in the pursuit of knowledge over these many years.

GFP

CONTENTS

PREFACE

This book introduces an innovative teaching methodology that seeks to give to students economy of effort in their learning, as well as clarity of vision and confidence in their knowledge. The strategy employed to achieve these purposes is an axiomatic approach. When an axiomatic approach is applied one identifies a set of properties (the 'set of axioms') such that all the other properties are concomitant to them. In this manner great generality is achieved, since any model that is formulated axiomatically is applicable to every system that satisfies its set of axioms and, depending on the choice of such axioms, the diversity of the systems covered by the model may be very wide.

Simplicity is also achieved when the axiomatic approach is used, because when identifying the set of axioms the most essential properties of the systems to be modeled must be singled-out and all those that are secondary or superfluous are eliminated. Furthermore, when the axioms are clearly stated the axiomatic approach also yields clarity of vision, which is essential for attaining confidence in the knowledge that is learned. It would be difficult to overstress the importance of developing such confidence when training scientists and engineers since it is the most effective manner of overcoming the limitations of the 'magister-dixit' (the 'master said it') paradigm of erstwhile educational methods.

The generality that is achieved when the axiomatic approach is followed has permitted us to cover in a one-semester course the basics of the most commonly used models in science and engineering. Treating such diverse topics typically requires several one-semester courses, when more traditional teaching methods are used. In the first three chapters of the book, the axiomatic formulation of continuous systems is introduced and used to derive models in their most basic form. Then, in the remaining chapters, the models are developed in a more detailed and extensive manner.

The specific topics of presented in this book are fundamental for a subject of outstanding relevance in contemporary science and engineering: predicting systems-behavior: Predicting the behavior of Nature and other systems of interest is not only an ancestral human longing but also a modern necessity of science and engineering. Nowadays, the method used for predicting systems-behavior is mathematical and computational modeling (MCM) which is applicable an enormous range of natural and man-made systems. In the MCM approach one constructs successively a conceptual model, a mathematical model, a numerical model and a computational model, which usually consists of a computational computer program or code. The conceptual model establishes the purpose for, and scope of the model to be developed; furthermore, it also identifies the processes and phenomena that will be incorporated in the mathematical model. Using numerical methods and matrix algebraic algorithms, this latter model is transformed into a numerical model. Then, a computational code is developed that permits solving the numerical equations with resource to suitable computational hardware.

This book deals with the second step of this modeling process, the mathematical modeling, for the case when the system to be modeled is a macroscopic physical one. Many systems of engineering, earth sciences, applied sciences and, indeed, sciences in general belong to this systems-category. In this book the axiomatic approach provides a systematic, yet simple, procedure for constructing the mathematical model (i.e., the basic system of differential equations) of any macroscopic-system, be it a physical, chemical or biological macroscopic-system.

ISMAEL HERRERA AND GEORGE F. PINDER

Mexico City, DF, Mexico and
Burlington, VT, USA
Winter 2012

CHAPTER 1

AXIOMATIC FORMULATION OF THE BASIC MODELS

1.1 MODELS

The general procedure for predicting the behavior of a system is *modeling*. In this process, models are built that are used to predict a system's behavior. A model of a system is a surrogate whose behavior mimics that of the system. Models come in two flavors: physically built and mathematical. The models used most extensively today are mathematical models, since they are the most versatile and inexpensive. In specific applications, mathematical models consist of computer codes that are easily adapted to accommodate changes in systems properties. Furthermore, the foundations and methods of application of mathematical models, as will be seen, possess remarkable conceptual unity and generality. These features have very important practical implications, because they lead to great economies of effort and resources.

Mathematical models integrate scientific and technological knowledge with the purpose of predicting system behavior. Such knowledge is, thereby, incorporated into the computational codes that the computers execute in model utilization. At present, mathematical models founded on numerical simulation permit the study of complex systems and natural phenomena that otherwise would be very costly, dangerous,

Mathematical Modeling in Science and Engineering: An Axiomatic Approach.
By Ismael Herrera and George F. Pinder Copyright © 2012 John Wiley & Sons, Inc.

or even impossible to study by direct experimentation. From this perspective the significance of mathematical and computational modeling is clear; it is the most efficient and effective method for predicting the behavior of both natural and artificial systems of human interest. Worldwide, mathematical models are used ,extensively in engineering and science, in endeavors as diverse as the oil industry, water supply, weather and climate prediction, automobile and aircraft manufacture, medical studies, economics, chemical studies, and astronomy.

1.2 MICROSCOPIC AND MACROSCOPIC PHYSICS

This book deals with models of *macroscopic physical systems*, which include many systems of engineering and science. For this purpose an axiomatic approach, which we discuss below, is used; this is the most effective procedure for achieving generality and conceptual unity [1-13]. In particular, the use of axioms will permit us to construct an efficient approach for getting acquainted with the most important models of science and engineering.

There exist two approaches to the study of matter and its motion: the *microscopic approach*, which studies molecules, atoms, and elemental particles; and the *macroscopic approach,* which studies and models large systems. Matter, when it is observed at the ultramicroscopic level, consists of molecules and atoms, which in turn are made up of even smaller particles, such as protons, neutrons, and electrons. Prediction of the behavior of ultramicroscopic particles is a microscopic approach and is the subject matter of *quantum mechanics.*

Predicting the behavior of very large systems, such as an oil reservoir or the atmosphere, which consist of exceedingly large numbers of molecules and atoms, is an inaccessible goal using the microscopic approach. Thus, for problems of practical importance in science and engineering, a macroscopic approach is the appropriate one. The *mechanics of continuous media* supplies the theoretical foundations of the latter approach. The basis for the axiomatic method applied to the mechanics of continuous media was established in the second half of the twentieth century by a group of scholars and researchers whose most conspicuous leaders were C. Truesdell and W. Noll [[17], [31]-[36]].

Many systems of interest in engineering, earth sciences, applied sciences and sciences in general require models based upon the mechanics of continuous media. Among them are models of civil engineering structures, soils and foundations, the Earth's crust and deep interior, blood flow, the mechanics of human bones, natural resource reservoirs such as those of oil and water (both, surface and groundwater), the atmosphere, climate and weather prediction, and many others.

To predict weather or climate by following the motion of each particle of the atmosphere is obviously impossible. However, by applying the concepts drawn from the mechanics of continuous media, computational and mathematical modeling has led to very impressive achievements and has treated a great diversity of problems of science and engineering. The theory of *macroscopic systems* can be applied not only to physical systems, but also to chemical and some biological systems as well. Using the continuum approach, for example, it is possible to predict the movement and evolution of microscopic biopopulations. In this approach, microscopic individuals are ignored; instead, the populations are assumed to be continuously distributed in all the space they occupy.

The study of this wide variety of problems in a unified manner by means of what we described earlier as the *axiomatic approach* yields an enormous economy of effort in the teaching-learning process. Furthermore, this manner of teaching and learning is very valuable in research, because the associated unified formulation contains clues for the solution of many heretofore unforeseen problems.

Some descriptions of the axiomatic approach to the formulation of continuous mechanics are difficult for non-mathematical audiences to follow, so herein we make an effort to simplify presentation of this approach without unduly compromising rigor. In particular, the presentation is similar to that contained in reference [1], which takes this approach. An important difference, however, is that we will employ the *intensive properties* by volume instead of by unit mass, since this yields significant advantages in the development and applications of the theory. We will now consider further the two approaches to viewing the world around us: that is, the *macroscopic and microscopic perspectives*.

When matter is studied using the macroscopic viewpoint, bodies completely fill the space they occupy, so that no voids exist in them, in spite of the fact that when they are examined with the help of a microscope, that is, from the microscopic point of view, one encounters many interstices. Thus, from the macroscopic point of view, for example, water fills the receptacle that contains it, and our work desk is a continuous piece of matter perfectly and sharply delimited.

This macroscopic viewpoint is present in classical physics, especially in classical mechanics. Science has now advanced to the point where we recognize that matter is full of voids that our senses do not perceive, and also that energy moves in subatomic packets; that is, it is *quantized*. At first glance these two approaches to the analysis of physical systems, the microscopic and the macroscopic, seem to be contradictory; however, they are not only compatible but are actually complementary, and a relation between them can be established by means of *statistical mechanics*.

1.3 KINEMATICS OF CONTINUOUS SYSTEMS

Let us begin with a consideration of some fundamental concepts. In the theory of continuous systems, as noted above, *material bodies completely fill the physical space that they occupy*. A *body* is a set of particles that, at any given time, occupies a *domain* (in the sense that this word is used in mathematics [[6]]) of the physical space. The body (that is, the set of particles) will be denoted by $\mathcal{B}$ and the domain that it occupies at time t will be denoted by $B(t)$. As for the time t, it can be any *real number*, that is, any number lying in the interval $-\infty < t < \infty$. However, in most studies of physical systems, the period of interest is contained in a finite time interval.

Given a body $\mathcal{B}$, other bodies that satisfy the condition $\mathcal{B} \supset B$ (that is, that are contained in it) will be said to be *subbodies of* $\mathcal{B}$.

Our discussion now proceeds by first considering *one-phase continuous systems* only, in which case a basic assumption of the theory is that given a body $\mathcal{B}$ at any time $t \in (-\infty, \infty)$, at each point $\underline{x} \in B(t)$[1] of the domain occupied by the body at such a time there is one and only one *particle* representative of or identified with that point in the body. For dynamical systems, particles change position as time evolves; thus a first problem with which the kinematics of continuous systems must deal is the identification of particles at different times. This is a challenge because generally the *same particle will have different locations at different times*.

A piece of information that identifies a particle uniquely is its position at a given time: the initial time, for example. Although this is not the only approach to identifying a particle, it is a very convenient one and one that will be used in what follows unless explicitly stated otherwise.

Let $p(\underline{X}, t)$ *be the particle's position vector at time* t; then if the particle X has been identified by means of its position $\underline{X}$ at the initial time ($t = 0$), the following identity is satisfied:

$$\underline{p}(\underline{X}, 0) \equiv \underline{X}. \tag{1.1}$$

The *vector coordinates* $\underline{X} \equiv (X_1, X_2, X_3)$ are referred to as the *material coordinates* of the particle, while the function $p(\underline{X}, t)$ is the *position-function*. We frequently reserve the notation $\underline{x}$ for the coordinates of the position of the particle in the physical space; then

$$\underline{x} = \underline{p}(\underline{X}, t). \tag{1.2}$$

The relationship presented in Eq. (1.2) is illustrated in Fig. 1.1.

Generally, $\underline{x} \neq \underline{X}$, unless $t = 0$, in which case Eq. (1.1) is fulfilled. Observe that Eq. (1.2) supplies an answer to the question: At time t, where is the position, in the physical space, of *particle* X? Another question that frequently occurs is: At time t, what particle is located at position $\underline{x}$ of the physical space? The answer to the latter question is

$$\underline{X} = \underline{p}^{-1}(\underline{x}, t). \tag{1.3}$$

At any given time, t, the function $p(\underline{X}, t)$ of Eq. (1.2) defines a transformation of a three-dimensional physical (Euclidean) space into itself. The notation $\underline{p}^{-1}(\underline{x}, t)$ is used for the *inverse transformation* of $p(\underline{X}, t)$. An example of the use of this transformation is presented in Fig. 1.2. The existence of $\underline{p}^{-1}(\underline{x}, t)$ is a basic axiom of continuum mechanics, sometimes referred to as the *axiom of bodies impenetrability*: For one-phase systems, two different particles cannot occupy the same position at any time. In particular, the trajectories of different particles cannot cross.

When the material coordinates are defined fulfilling Eq. (1.1), then the set $\mathcal{B}$ is the domain that the body occupies at the initial time, and $\underline{X} \in B(0)$ if and only if the particle belongs to the body. On the other hand, as time passes the position in the

---

[1] Note that an underline denotes a vector quantity.

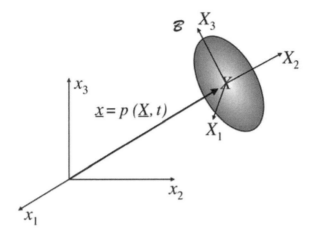

Figure 1.1 Relationship between material coordinates $\underline{X}$ and the coordinates of the position of the particle X in physical space $\underline{x}$.

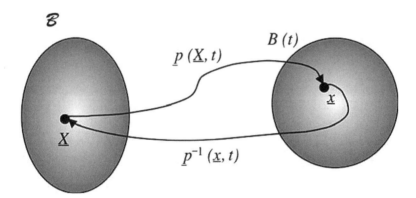

Figure 1.2 Relationship between the forward problem $x = \underline{p}\underline{X}, t)$ and the inverse relationship $\underline{X} = \underline{p}^{-1}(\underline{x}, t)$.

physical space of the body changes and $B(t)$ is the domain occupied by the body at any time t (see Fig. 1.2). In particular,

$$B = B(0). \tag{1.4}$$

A formal definition of $B(t)$ is:

$$B(t) \equiv \left\{ \underline{x} \in R^3 \middle| \exists \underline{X} \in B \ni \underline{x} = \underline{p}(\underline{X}, t) \right\} \tag{1.5}$$

which reads \a body B that is a function of time " is defined as follows: $\underline{x}$ is an element of R^3 such that there exists an $\underline{X}$ that is an element of B such that $\underline{x}$ is defined by the relationship $\underline{x} = \underline{p}(\underline{X}, t)$.

If a particle $\underline{X}$ is kept fixed while the time is varied, the function $\underline{p}(\underline{X}, t)$ yields the *trajectory of the particle*. In particular, the *velocity of a particle* is the derivative with respect to time of the function $\underline{p}(\underline{X}, t)$, when the choice of particle X is kept fixed. Thus, we define the *particle velocity* by

$$\underline{V}(\underline{X}, t) \equiv \frac{\partial \underline{p}}{\partial t}(\underline{X}, t). \tag{1.6}$$

1.3.1 Intensive properties

In continuum mechanics there is a class of functions, such as density or temperature, which have to be studied that are defined for each *particle* of a *body* and for each time. In general, any such function will be said to be an *intensive property*. When we consider an intensive property, we will write $\phi(\underline{X}, t)$ for its value at the particle location $\underline{X}$ at time t. This function, $\phi(\underline{X}, t)$, is said to be the *Lagrangian representation* of the intensive property.

On the other hand, one can consider this same intensive property from another point of view; let $\psi(\underline{x}, t)$ be the value of the intensive property at point $\underline{x}$ of the physical space at time t. The function $\psi(\underline{x}, t)$ is said to be the *Eulerian representation* of the intensive property. There is a condition that the Lagrangian and Eulerian representations of the same intensive property have to satisfy; it follows from Eq. (1.2) and is

$$\phi(\underline{X}, t) \equiv \psi(\underline{p}(\underline{X}, t), t) = \psi(\underline{x}, t). \tag{1.7}$$

In particular, the functions $\phi(\underline{X}, t)$ and $\psi(\underline{x}, t)$ are not the same. Furthermore, Eq. (1.7) implies that

$$\psi(\underline{x}, t) \equiv \phi(\underline{p}^{-1}(\underline{x}, t), t). \tag{1.8}$$

Clearly, Eqs. (1.7) and (1.8) are equivalent. The names we have adopted for these representations of intensive properties are to honor the mathematicians Leonard Euler (1707-1783) and Joseph-Louis Lagrange (1736-1813), respectively. The Lagrangian representation is used more frequently in the study of solids, while the Eulerian representation is more often used in the study of fluids. This is probably due to the fact that fluid displacements are usually very large, while displacements of solids are generally small.

Intensive properties may be scalar functions or vector functions. As an example, the particle velocity defined by Eq. (1.6) is a *vector-intensive property*. We notice that Eq. (1.6) yields the Lagrangian representation of the particle velocity. Also observe that a vector-intensive property is equivalent to three *scalar-intensive properties*: one scalar-intensive property for each component of the particle velocity.

In view of Eq. (1.7), the Lagrangian and Eulerian representations of the particle velocity satisfy

$$\underline{V}(\underline{X}, t) \equiv \underline{v}(\underline{p}(\underline{X}, t), t). \tag{1.9}$$

The Eulerian representation of the particle velocity, here denoted by $\underline{v}(\underline{x}, t)$, in view of Eq. (1.8) is given by

$$\underline{v}(\underline{x}, t) \equiv \underline{V}(\underline{p}^{-1}(\underline{x}, t), t). \tag{1.10}$$

The partial derivative with respect to time, $\partial\psi(\underline{x}, t)/\partial t$, of the Eulerian representation of an intensive property is simply the rate of change of that property at a point that *remains fixed in the physical space*. On the other hand, the partial derivative with respect to time of the Lagrangian representation, $\partial\phi(\underline{X}, t)/\partial t$, is, according to the definition of a partial derivative, the rate of change of such a property *that takes place at the particle under consideration as it moves in the space*; therefore, it is the rate of change that takes place at a point that moves in the physical space with the particle velocity. Or, to put this concept in more intuitive terms, it is the rate of change that we would observe if we rode on the particle and measured the intensive property.

It is useful to supply a formula for evaluating $\partial\phi(\underline{X}, t)/\partial t$ in terms of the Eulerian representation, $\psi(\underline{x}, t)$, of the property. Such a relationship can easily be derived from Eq. (1.7). Indeed, if we differentiate Eq. (1.7) with respect to time using the *chain rule*, we obtain, using Eq. (1.2),

$$\frac{\partial\phi(\underline{X}, t)}{\partial t} = \frac{\partial\psi}{\partial t}(\underline{p}(\underline{X}, t), t) + \sum_{i=1}^{3} \frac{\partial\psi}{\partial x_i}(\underline{p}(\underline{X}, t), t)\frac{\partial p_i}{\partial t}(\underline{X}, t) \tag{1.11}$$

or,

$$\frac{\partial\phi(\underline{X}, t)}{\partial t} = \frac{\partial\psi}{\partial t}(\underline{p}(\underline{X}, t), t) + \sum_{i=1}^{3} \frac{\partial\psi}{\partial x_i}(\underline{p}(\underline{X}, t), t)V_i(\underline{X}, t). \tag{1.12}$$

Recalling Eq. (1.9) written in component form,

$$V_i(\underline{X}, t) = v_i(\underline{p}(\underline{X}, t), t). \tag{1.13}$$

Eq. (1.12) becomes

$$\frac{\partial\phi(\underline{X}, t)}{\partial t} = \frac{\partial\psi}{\partial t}(\underline{p}(\underline{X}, t), t) + \sum_{i=1}^{3} \frac{\partial\psi}{\partial x_i}(\underline{p}(\underline{X}, t), t)v_i(\underline{p}(\underline{X}, t), t) \tag{1.14}$$

or, more briefly,

$$\frac{\partial\phi(\underline{X}, t)}{\partial t} = \frac{\partial\psi}{\partial t}(\underline{x}, t) + \underline{v}(\underline{x}, t) \cdot \boldsymbol{\nabla}\psi(\underline{x}, t). \tag{1.15}$$

We will use the symbol $D\psi/Dt$ for the time derivative of the *Lagrangian representation*; it is given by

$$D\psi/Dt = \partial\psi/\partial t + \underline{v} \cdot \boldsymbol{\nabla}\psi. \tag{1.16}$$

In particular, the acceleration of a *material particle, $\underline{a}(\underline{x}, t)$,* is given by

$$\underline{a}(\underline{x}, t) = (\partial\underline{v}/\partial t + \underline{v} \cdot \boldsymbol{\nabla}\underline{v})(\underline{x}, t) \tag{1.17}$$

or

$$a_i(\underline{x}, t) = (\partial v_i/\partial t + \underline{v} \cdot \boldsymbol{\nabla}v_i)(\underline{x}, t); \ i = 1, ..., 3. \tag{1.18}$$

It follows from Eq. (1.6) that the Lagrangian representation of the acceleration is simply

$$\frac{\partial}{\partial t}V(\underline{X}, t) \equiv \frac{\partial^2}{\partial t^2}\underline{p}(\underline{X}, t). \tag{1.19}$$

1.3.2 Extensive properties

In this section we consider functions (that is, properties) that are *defined for each body of a continuous system*. This is to be contrasted with the functions (or properties) that we considered in the preceding section, which were defined for each material particle. We start with the definition of an extensive property: $E(\mathcal{B}, t)$ is said to be an *extensive property* when for every $\mathcal{B}$ it can be expressed as an integral of an intensive property over the body ; that is,

$$E(\mathcal{B}, t) \equiv \int_{B(t)} \psi(\underline{x}, t)\, d\underline{x}. \tag{1.20}$$

With respect to this definition, we note the following. Firstly, it is assumed that the Eulerian representations, $\psi(\underline{x}, t)$, of intensive properties to be considered in this section are integrable in the sense of *Lebesgue*.[2] Then, Eq. (1.20) establishes a one-to-one correspondence between extensive properties and intensive properties because the integrand function in Eq. (1.20) can always be taken as the definition of the Eulerian representation of an intensive property. Conversely, given an intensive property, its integral over each body defines an extensive property.

Second, the intensive property associated in this manner with each extensive property is the extensive property *per unit volume*. Note that in classical approaches (see, for example, reference [1]) intensive properties *per unit mass* have been used instead. However, the use of intensive properties per unit volume permits achieving greater theoretical consistency and more elegant developments. In particular, the correspondence between intensive and extensive properties is more direct. For example, when an extensive property is given by Eq. (1.20), its corresponding intensive property is the integrand in that equation.

In addition, when dealing with complicated systems such as multiphase and multicomponent systems, many masses may be involved in the systems and the application of formulas which use properties defined per unit mass may be confusing. On the other hand, the volume of the physical space is always uniquely defined.

It is also noteworthy that the computations required for transformation between the two definitions of intensive properties are easy to do: *The intensive property per unit volume equals the intensive property per unit mass multiplied by the mass density.*

Thus a paradox that the use of intensive properties per unit mass leads to is identified: When intensive properties are defined as the associated extensive property per unit mass and we try to be rigorous and systematic in the use of such a concept, the intensive property associated with mass (as an *extensive property*) is unity (so it is not density). Finally, calculus shows that

$$\psi(\underline{x}, t) \equiv \lim_{V \to 0} \frac{E(t)}{V} = \lim_{V \to 0} \frac{\int_{B(t)} \psi(\underline{\xi}, t)\, d\underline{\xi}}{V}. \tag{1.21}$$

---

[2]The main advantage of the Lebesgue integral approach over the more common Riemann integral approach is that more functions become integrable.

Equation (1.21) supplies an effective means for experimentally determining intensive properties (in the sense in which we have defined them and will be using them in what follows) associated with an extensive one: When the volume of the sample, V, is sufficiently small, the intensive property equals, approximately, the quotient of the extensive property divided by the volume of the sample.

A comment on notation: In Eq. (1.20) we have written $E(\mathcal{B}, t)$ for the extensive property, to emphasize that the extensive property is a function of the body $\mathcal{B}$ considered. However, to simplify the notation, from now on we will write simply $E(t)$, dropping $\mathcal{B}$, except in situations in which such a practice may create ambiguity.

On the other hand, we will use the concepts of intensive and extensive properties with considerable freedom, but always in a logically consistent manner. In particular, we will consider intensive and extensive properties as those that satisfy the conditions stated in the definitions of these concepts. It must be pointed out, however, that not every property that we may think of which satisfies such definitions will be physically relevant. That notwithstanding, the concepts of intensive and extensive properties are fundamental in the development of *continuum mechanics*, mainly because the basic models of continuous systems consist of balance equations of such properties.

1.4   BALANCE EQUATIONS OF EXTENSIVE AND INTENSIVE PROPERTIES

As said above, the basic mathematical models of continuous systems are formulated via balance equations operating on certain families of extensive properties. As an example, the models of solute transport (and the transport of contaminants by the atmosphere or by water in the surface or the subsurface are particular cases of this general kind of problem) are generated by carrying out the balance of the solute mass contained in every domain of the physical space. Here, the word *balance* is used in essentially the same sense as it is used in accounting. In accounting, the income minus the expenditure equals the net change in capital. Similarly, in continuum mechanics, the flux into minus the flux out of an extensive property equals its net change in this property.

1.4.1   Global balance equations

In order to calculate such balances it is necessary, first, to carry out an exhaustive identification of the possible causes by which the extensive property that is being considered may change. Let us take as an example the calculation of the balance of the number of automobiles that exist in a country. A simple analysis shows that there are only four possible causes for a change in such a number: Either automobiles are produced or destroyed in the interior of the country or they are imported or exported through its boundary (borders). In formulating our balance equation the concepts of producing, destroying, importing, and exporting must be used in a *mathematical* sense; that is, destroying (or junking) is *negative production*, while exporting is *negative importation*. Given this formalism, these two concepts completely cover all possibilities of change. The balance equation becomes: *The net change in the number of automobiles in the country is due to the net production plus the net importation.*

To establish the balance equation of an extensive property, E, we will essentially follow the same steps as outlined above; the change in the extensive property, ΔE, will be given by

$$\Delta E = P + I. \tag{1.22}$$

Here, we have written P for the *production* of the extensive property in the interior of a domain of the physical space, and I for the *importation* of it through the boundary of such a domain.

Recalling that the *material particles* contained in a domain of the physical space constitute a body, we can use the ideas above to establish the balance equations that an extensive property of a body must fulfill. If we now introduce the concept of time, that is, the change in property E per unit time, we can use Eq. (1.22) to obtain

$$\frac{dE}{dt}(t) = \int_{B(t)} g\left(\underline{x}, t\right) d\underline{x} + \int_{\partial B(t)} q\left(\underline{x}, t\right) d\underline{x}. \tag{1.23}$$

The first integral is taken over the volume of the domain $B\left(t\right)$ and the second is taken over the surface $\partial B\left(t\right)$ of the domain $B\left(t\right)$. Here $g\left(\underline{x}, t\right)$ is called the *external supply of the extensive property, per unit volume and per unit time*; it represents the amount of the extensive property that enters the body, at point $\underline{x}$ and time t, per unit volume. As for $q\left(\underline{x}, t\right)$, *it represents the amount of the extensive property that enters the body through its boundary, at point $\underline{x}$ and time t, per unit area.* It can be shown, under very general conditions, that $q\left(\underline{x}, t\right)$ is given by

$$q\left(\underline{x}, t\right) \equiv \underline{\tau}\left(\underline{x}, t\right) \cdot \underline{n}\left(\underline{x}, t\right). \tag{1.24}$$

Here, $\underline{n}\left(\underline{x}, t\right)$ is the *unit normal vector* on $\partial B\left(t\right)$, pointing toward the exterior of the body. The vector $\underline{\tau}$ is the *flux* of the *extensive property*. Using this relation, the balance represented by Eq. (1.23) can be rewritten as

$$\frac{dE}{dt}(t) = \int_{B(t)} g\left(\underline{x}, t\right) d\underline{x} + \int_{\partial B(t)} \underline{\tau}\left(\underline{x}, t\right) \cdot \underline{n}\left(\underline{x}, t\right) d\underline{x}. \tag{1.25}$$

This equation states that the rate of increase of property E in the body $B\left(t\right)$ is equal to the rate of introduction of property E introduced into the body $B(t)$ via an external supply plus the amount per unit time entering through the boundary $\partial B\left(t\right)$. This relationship is generally referred to as the *general equation of global balance*, which is a fundamental concept in the formulation of mathematical models of continuous systems.

1.4.2 The local balance equations

In this section we derive the *local balance equations*, which are expressed in terms of the intensive properties of continuous systems. The notation $\Sigma\left(t\right)$ will be used to denote a *shock*: that is, the surface across which the intensive properties are discontinuous. In this notation the time t is included to make explicit that the shock position in the physical space generally depends on the time; that is, the shock generally moves. Standing shocks are included as particular cases of our general framework, just as the notation $f(\underline{x})$ in which the dependence on the argument $\underline{x}$ is made explicit includes as particular cases the constant functions.

The kind of discontinuities that may occur across $\Sigma(t)$ are jump discontinuities. We recall that, by definition, a *jump discontinuity* of a function is one in which the limits from each side of $\Sigma(t)$ exist but are different from each other. On $\Sigma(t)$ we define a *positive side* arbitrarily and then a unit normal vector pointing towards the positive side. Given a function, $f(\underline{x})$, with a jump discontinuity across $\Sigma(t)$, we define its *jump* and its *average*, respectively, by

$$[\![f]\!] \equiv f_+ - f_- \text{ and } \widehat{f} \equiv \frac{1}{2}(f_+ + f_-). \tag{1.26}$$

Here f_+ and f_- stand for the limits of f on the positive and negative sides of $\Sigma(t)$, respectively.

Furthermore, due to the presence of *jump discontinuities* on the shock, concentrated sources also occur frequently on $\Sigma(t)$. Due to this fact Herrera introduced a more general form of the *equation of global balance* [[13],[15],[16]] of Eq. (1.25), that will be used throughout this book. It is

$$\frac{dE}{dt}(t) = \int_{B(t)} g(\underline{x}, t)\, d\underline{x} + \int_{\partial B(t)} \underline{\tau}(\underline{x}, t) \cdot \underline{n}(\underline{x}, t)\, d\underline{x} + \int_{\Sigma(t)} g_\Sigma(\underline{x}, t)\, d\underline{x}. \tag{1.27}$$

Here $g_\Sigma(\underline{x}, t)$ represents an *external supply* that is concentrated on Σ.

The following lemma will be used later in our development.

Lemma 1 *-The general global balance equation, Eq.(1.27), is equivalent to*

$$\frac{dE(t)}{dt} = \int_{B(t)} \{g(\underline{x}, t) + \boldsymbol{\nabla}\!\cdot\!\underline{\tau}(\underline{x}, t)\}\, d\underline{x} + \int_{\Sigma(t)} \{[\![\underline{\tau}]\!] \cdot \underline{n}(\underline{x}, t) + g_\Sigma(\underline{x}, t)\}\, d\underline{x}. \tag{1.28}$$

Proof: Under the assumption that the flux, $\underline{\tau}$, is piecewise continuous with jump discontinuities on $\Sigma(t)$, exclusively, Eq.(1.28) is derived by a straight-forward application of the generalized Gauss theorem described in the Appendix B.   ∎

Next we present a mathematical result that also will be used later.

Theorem 2 *For each real number t, let $B(t) \subset R^3$ be a domain in which a body is located. Let $\psi(\underline{x}, t)$ be a piecewise C^1 continuous intensive property [that is, C^1 except across $\Sigma(t)$]). Furthermore, let $\underline{v}(\underline{x}, t)$ and $\underline{v}_\Sigma(\underline{x}, t)$ be the particle velocity and the surface velocity of $\Sigma(t)$, respectively. Then*

$$\frac{d}{dt}\int_{B(t)} \psi\, d\underline{x} \equiv \int_{B(t)} \left\{\frac{\partial \psi}{\partial t} + \boldsymbol{\nabla}\cdot(\underline{v}\psi)\right\} d\underline{x} + \int_{\Sigma(t)} [\![(\underline{v} - \underline{v}_\Sigma)\,\psi]\!] \cdot \underline{n}\, d\underline{x} \tag{1.29}$$

Proof: A proof is given in Appendix C. ∎

Corollary 3 *Let* $\psi\left(\underline{x}, t\right)$ *be the intensive property that corresponds to the extensive property* $E\left(t\right)$. *Then*

$$\frac{dE}{dt} \equiv \int_{B(t)} \left\{ \frac{\partial\psi}{\partial t} + \boldsymbol{\nabla}\cdot\left(\underline{v}\psi\right) \right\} d\underline{x} + \int_{\Sigma(t)} \left[\left[\left(\underline{v}-\underline{v}_{\Sigma}\right)\psi\right]\right]\cdot\underline{n}\, d\underline{x} \qquad (1.30)$$

Proof: Equations (1.20) *and* (1.26) *together imply Eq.* (1.30). ∎

Corollary 4 *The general global balance equation, Eq. 1.27, is fulfilled, if and only if, for every body* $B(t)$ *the following relationship holds:*

$$\int_{B(t)} \left\{ \frac{\partial\psi}{\partial t} + \boldsymbol{\nabla}\cdot\left(\underline{v}\psi\right) - \boldsymbol{\nabla}\cdot\underline{\tau}\left(\underline{x}, t\right) - g\left(\underline{x}, t\right) \right\} d\underline{x}$$

$$+ \int_{\Sigma(t)} \left\{ \left[\left[\left(\underline{v}-\underline{v}_{\Sigma}\right) - \underline{\tau}\right]\right]\cdot\underline{n}\left(\underline{x}, t\right) - g_{\Sigma}\left(\underline{x}, t\right) \right\} d\underline{x} = 0. \qquad (1.31)$$

Proof: Equations. (1.28) *and 1.30 together imply that*

$$\int_{B(t)} \left\{ \frac{\partial\psi}{\partial t} + \boldsymbol{\nabla}\cdot\left(\underline{v}\psi\right) \right\} d\underline{x} + \int_{\Sigma(t)} \left[\left[\left(\underline{v}-\underline{v}_{\Sigma}\right)\psi\right]\right]\cdot\underline{n}\left(\underline{x}, t\right) d\underline{x}$$

$$= \int_{B(t)} \left\{ g\left(\underline{x}, t\right) + \boldsymbol{\nabla}\cdot\underline{\tau}\left(\underline{x}, t\right) \right\} d\underline{x} + \int_{\Sigma(t)} \left\{ \left[\left[\underline{\tau}\right]\right]\cdot\underline{n}\left(\underline{x}, t\right) + g_{\Sigma} \right\} d\underline{x}.$$

∎

Theorem 5 *- Let* $B\left(t\right)$ *be a body. Then the global balance equation, Eq.(1.27), is fulfilled at every sub-body contained in* $B\left(t\right)$ *if, and only if, the following conditions are satisfied:*

1. The differential equation

$$\frac{\partial\psi}{\partial t} + \boldsymbol{\nabla}\cdot\left(\underline{v}\psi\right) = \boldsymbol{\nabla}\cdot\underline{\tau} + g,\ \text{holds at every } \underline{x} \in B\left(t\right) - \Sigma\left(t\right). \qquad (1.32)$$

2. The jump condition

$$\left[\left[\psi\left(\underline{v}-\underline{v}_{\Sigma}\right) - \underline{\tau}\right]\right]\cdot\underline{n} = g_{\Sigma},\ \text{holds at every } \underline{x} \in \Sigma\left(t\right) \qquad (1.33)$$

Proof: That Eqs.(1.32) and (1.33) imply Eq.(1.31) is obvious. On the other hand, Eqs.(1.32) and (1.33) can be derived from the fact that Eq.(1.31) must be fulfilled for every subbody of $B\left(t\right)$. ∎

The details of the proof of the latter result are left as an exercise at the end of this chapter.

In what follows, Eq. (1.32) will be referred to as the *differential equation of local balance*, while Eq. (1.33) will be the jump condition. It is timely to mention that the differential equation of local balance is used more extensively than the jump condition. Indeed, practically all models of continuous media are based on Eq. (1.32), while the jump condition is used only when the model considered contains surfaces of discontinuity, and the latter situation seldom occurs. Nevertheless, a thorough discussion of continuous systems requires the study of the jump condition.

It is also useful to be acquainted with alternative expressions for the differential equation of local balance that are frequently used. In particular, each of the equations

$$\frac{\partial \psi}{\partial t} + \underline{v} \cdot \boldsymbol{\nabla} \psi + \psi \boldsymbol{\nabla} \cdot \underline{v} = \boldsymbol{\nabla} \cdot \underline{\tau} + g \tag{1.34}$$

and

$$\frac{D\psi}{Dt} + \psi \boldsymbol{\nabla} \cdot \underline{v} = \boldsymbol{\nabla} \cdot \underline{\tau} + g \tag{1.35}$$

is equivalent to Eq. (1.32). For *stationary states* of continuous systems $\partial \psi / \partial t = 0$, Eq. (1.32) reduces to

$$\boldsymbol{\nabla} \cdot (\underline{v}\psi) = \boldsymbol{\nabla} \cdot \underline{\tau} + g. \tag{1.36}$$

Then, the jump condition presented in Eq. (1.33) becomes

$$[\![\psi \underline{v} - \underline{\tau}]\!] \cdot \underline{n} = g_\Sigma \text{ on } \Sigma\,(t) \tag{1.37}$$

since for stationary states one usually has $\underline{v}_\Sigma \equiv 0$.

1.4.3 The role of balance conditions in the modeling of continuous systems

As will be seen, associated with each model of a continuous system there is a unique family of extensive properties. Then the basic mathematical model of such a continuous system is constituted by the family of balance equations corresponding to each extensive property of the associated family, whose global formulations are given by Eq. (1.27). However, extensive properties are not used directly to express the local balance conditions; instead, such balances at the local scale are expressed in terms of their intensive properties. This substitution of extensive properties by their corresponding intensive properties is feasible because the global balance equation of Eq. (1.27) is equivalent to the local balance equations of Eqs. (1.32) and (1.33), and the latter do not involve extensive properties but rather their associated intensive properties.

The applicability of the local balance equations of Eqs. (1.32) and (1.33) is very wide since it includes models of continuous systems that contain surfaces of discontinuity. As noted earlier, such surfaces of discontinuity are frequently called *shocks*. Well-known examples of shocks are those that occur in supersonic flow of non-viscous compressible fluids. Other examples occur in hydraulics and in oil-reservoir mechanics. Shocks model very rapid changes that take place in the physical systems.

We observe that the local balance equations involve two kinds of relations that must be satisfied by the intensive properties: the balance differential equations and the jump conditions; the former are differential equations that are fulfilled at every interior point of a body, while the latter are conditions to be fulfilled at the surfaces of discontinuity, or shocks. The use of Eq. (1.33) permits treating the shock phenomenon in a unified manner.

The balance differential equations, together with the jump conditions, constitute a *generic mathematical model*. In order for a generic model to acquire the capacity to predict behavior, it is necessary to incorporate in the model scientific and technological knowledge about the specific systems under consideration; this is achieved by means of *constitutive equations*. Once the constitutive equations have been combined with the balance differential equation, and the jump conditions, the problem of developing the model of the continuous system and deriving predictions from it becomes one that belongs to the fields of *partial differential equations* and *numerical methods*. Scientific computation, as used herein, supplies the methods and techniques required for applying the resulting mathematical model to specific problems of engineering and science.

1.4.4   Formulation of motion restrictions by means of balance equations

As a first application of the balance equations, in this subsection we formulate the conditions of *incompressibility of a fluid*, in two physical situations of practical interest: when the fluid is free to move in space and when it is contained in a porous medium.

The movement of a continuous system is said to be *isochoric* when every body of such a system conserves its volume. Clearly, the volume of a body is given by

$$V_B(t) \equiv \int_{B(t)} d\underline{x} = \int_{B(t)} 1 \, d\underline{x}. \tag{1.38}$$

When the movement is isochoric, one has

$$\frac{dV_B}{dt}(t) = 0. \tag{1.39}$$

Comparing Eqs. (1.38) and (1.39) with Eqs. (1.20) and (1.25), it is seen that in this case $\psi = 1$, as well as $g = 0$ and $\underline{\tau} = 0$. Substituting these values in the local balance differential equation, Eq. (1.32), one obtains

$$\nabla \cdot \underline{v} = 0. \tag{1.40}$$

This is the well-known *condition of incompressibility* that applies to a free fluid.

Assume now that we consider a fluid restricted to move in the pores, $V_v(t)$, of a porous medium (Fig. 1.3) and the question we would like to answer is: For a fluid contained in this porous medium, what are the conditions that the velocity satisfies when the motion is isochoric? The answer to this question is not as well known as that for a free fluid, but the question is answered as easily using the general scheme that we have already established. Consider the *volume of the pores* of the porous medium. This is an extensive property because it can be written as an integral over the domain where the porous medium is located:

$$V_p(t) = \int_{B(t)} \varepsilon(\underline{x}, t)\, dx. \tag{1.41}$$

The integrand in this expression, $\varepsilon(\underline{x}, t)$, is the associated intensive property, known as *porosity*.It can be obtained in the laboratory by means of Eq. (1.21), which in this case yields

$$\varepsilon(\underline{x}, t) = \frac{\text{volume of voids}}{\text{total volume}}. \tag{1.42}$$

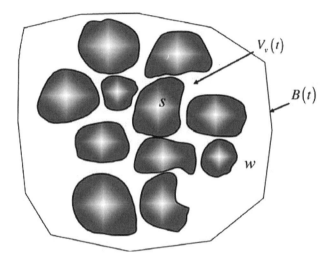

Figure 1.3   Diagrammatic representation of a porous medium. The medium is made up of solid s and water w.

A porous medium is said to be *saturated* by a fluid, w, when the pores are completely full of water; that is, when the volume of the fluid is equal to the volume of the pores. For isochoric motion of the fluid we can apply the general global balance equation, Eq. (1.27), taking as the extensive property the volume of the fluid (equal to the volume of the pores) with $g = 0$ and $\underline{\tau} = 0$ and $g_\Sigma = 0$. Therefore, the differential equation of local balance, Eq. (1.32), is

$$\frac{\partial \varepsilon}{\partial t} + \nabla \cdot (\varepsilon \underline{v}) = 0. \tag{1.43}$$

This equation is the *condition of incompressibility for a fluid contained in a porous medium*. Its physical meaning and application will be discussed in Chapter 5.

In subsurface hydrology and petroleum engineering the discontinuities of the porosity are most frequently associated with surfaces across which two different geological strata are in contact. Such surfaces are fixed in the physical space and, therefore, $\underline{v}_\Sigma = 0$. Applying the jump conditions of Eq. (1.33), again with $\underline{\tau} = 0$ and $\psi = \varepsilon$, we get

$$[\![\varepsilon \underline{v}]\!] \cdot \underline{n} = 0 \text{ on } \Sigma. \tag{1.44}$$

This is the condition that the jump of the particle velocity of a fluid has to satisfy when the fluid goes from one geologic layer into another. The proof of the following algebraic identity is left as an exercise:

$$[\![\varepsilon \underline{v}]\!] = [[\varepsilon \underline{v}]] = [[\varepsilon]]\,\widehat{\underline{\dot{v}}} + \widehat{\dot{\varepsilon}}\,[[\underline{v}]]. \tag{1.45}$$

Using it, and Eq. (1.44), it is seen that

$$[[\underline{v} \cdot \underline{n}]] = -\frac{[[\varepsilon]]}{\widehat{\dot{\varepsilon}}}\,\widehat{\underline{\dot{v}}} \cdot \underline{n}. \tag{1.46}$$

Therefore: *Whenever the porosity has a surface of discontinuity, the normal component of the particle velocity also has a discontinuity, which is proportional to its average.* The commonly used *Darcy velocity* is defined as

$$\underline{U} \equiv \varepsilon \underline{v}. \tag{1.47}$$

Using it, the jump condition of Eq. (1.44) can be written as

$$[\![\underline{U}]\!] \cdot \underline{n} = 0 \text{ on } \Sigma. \tag{1.48}$$

That is, the jump condition requires that the normal component of the Darcy velocity be continuous across surfaces of discontinuity of the porosity, albeit the normal component of the particle velocity is discontinuous there.

1.5 SUMMARY

Following a brief introduction to the scope of this book and the concept of modeling, we introduced several ideas that are central to understanding the material presented in later chapters. Representation of physical systems from microscopic and macroscopic perspectives was presented, followed by a consideration of the kinematics of continuous systems. An important aspect of this discussion was the identification of Eulerian and Lagrangian coordinate systems and their relationship to one another. The concepts of intensive and extensive properties were next presented and used to derive two equivalent formulations of the balance equations: one in terms of extensive properties (the global balance conditions) and the other in terms of intensive properties (the global local balance conditions). The chapter concluded with an application of the balance equations to formulate constraints in the motion of continuous systems.

EXERCISES

In each of the following two exercises, the motions of one-phase continuous systems are given by means of corresponding position functions $\underline{p}(\underline{X}, t)$. It is required to determine:

a. The initial particle positions

b. The function $\underline{p}^{-1}(\underline{x}, t)$

c. The Lagrangian representation of the particle velocity, $\underline{V}(\underline{X}, t)$

d. The Eulerian representation of the particle-velocity, $\underline{v}(\underline{x}, t)$

e. The Jacobian matrix of the motion:

$$\underline{\underline{J}}(t) \equiv \frac{\partial p_i}{\partial X_j}(\underline{X}, t) \tag{1.49}$$

1.1 The motion is given by

$$
\begin{aligned}
p_1(\underline{X}, t) &\equiv X_1 + t \\
p_2(\underline{X}, t) &\equiv X_2 + 3t \\
p_3(\underline{X}, t) &\equiv X_3 + te^{-t}.
\end{aligned} \tag{1.50}
$$

1.2 The motion is given by

$$\underline{p}(\underline{X}, t) \equiv \left(\|\underline{X}\|^2 + 3\,\|\underline{X}\|^{-1}\,t \right)^{\frac{1}{3}} \underline{X}. \tag{1.51}$$

1.3 The Lagrangian representation of the particle velocity of a one-phase continuous system is given by

$$
\begin{aligned}
V_1(\underline{X}, t) &= 3t \\
V_2(\underline{X}, t) &= \sin(2t) \\
V_3(\underline{X}, t) &= X_3.
\end{aligned} \tag{1.52}
$$

Assuming that the initial particle positions are taken as *material coordinates,* it is required to obtain

    a) The *position function,* $\underline{p}(\underline{X}, t)$

    b) The Eulerian representation of the *particle velocity,* $\underline{v}(\underline{x}, t)$

    c) For the particle $\underline{X} = (1, 1, 0)$, a graphic of its trajectory in the time interval $0 \leq t \leq 1$

1.4 Let $\underline{p}(\underline{X}, t)$ be the position-function. The Lagrangian representation of the velocity is given by $V(\underline{X}, t) \equiv \frac{\partial p}{\partial t}(\underline{X}, t)$. Furthermore, let $\underline{v}(\underline{x}, t)$ and $\underline{a}(\underline{x}, t)$

be the Eulerian representations of the velocity and acceleration, respectively. It is required to establish:

a) The relation between $\underline{v}(\underline{x}, t)$ and $V(\underline{X}, t)$

b) The proof of the equation

$$\underline{a}(\underline{x}, t) = \frac{\partial \underline{v}}{\partial t}(\underline{x}, t) + \frac{1}{2}\nabla(v^2)(\underline{x}, t) - \underline{v}(\underline{x}, t) \times (\nabla \times \underline{v})(\underline{x}, t) . \tag{1.53}$$

1.5 Let u and v be two piecewise-defined functions and Σ a surface where they have jump discontinuities. Prove that

$$[[uv]] \equiv (u_+ v_+ - u_- v_-) = \dot{u}\,[[v]] + \dot{v}\,[[u]] . \tag{1.54}$$

Here

$$[[u]] \equiv (u_+ - u_-) \text{ and } \dot{u} \equiv \frac{1}{2}(u_+ + u_-) \tag{1.55}$$

and similarly for v. Using these relations, show Eq. (1.47). The equation

$$\frac{dE}{dt}(t) = \int_{B(t)} \left\{ \frac{\partial \psi}{\partial t} + \nabla \cdot (\psi \underline{v}) \right\} d\underline{x} \tag{1.56}$$

is a mathematical identity; when the domain of an integral depends on a parameter, t, it supplies an expression for the derivative with respect to such a parameter. It can be applied when

$$E(t) \equiv \int_{B(t)} \psi \, d\underline{x}. \tag{1.57}$$

A more intuitive form of this equation is

$$\frac{dE}{dt}(t) = \int_{B(t)} \frac{\partial \psi}{\partial t} d\underline{x} + \int_{\partial B(t)} \psi \underline{v} \cdot \underline{n} \, d\underline{x} \tag{1.58}$$

Prove that Eqs. (1.56) and (1.58) are equivalent. Furthermore, draw a figure to supply an intuitive interpretation of this result. To be valid , Eq. (1.51) requires that the function ψ be continuous. Show that when ψ is piecewise continuous, Eq. (1.51) must be replaced by

$$\frac{dE}{dt}(t) = \int_{B(t)} \frac{\partial \psi}{\partial t} d\underline{x} + \int_{\partial B(t)} \psi \underline{v} \cdot \underline{n} \, d\underline{x} - \int_{\Sigma(t)} [[\psi \underline{v}]] \cdot \underline{n} \, d\underline{x} \tag{1.59}$$

1.6 Show that the following expressions for the differential balance equations are equivalent:

$$\frac{\partial \psi}{\partial t} + \nabla \cdot (\underline{v}\psi) = g + \nabla \cdot \underline{\tau} \tag{1.60}$$

$$\frac{\partial \psi}{\partial t} + \frac{\partial}{\partial x_j}(v_j \psi) = g + \frac{\partial \tau_j}{\partial x_j} \tag{1.61}$$

and

$$\frac{D\psi}{Dt} + \psi \nabla \cdot \underline{v} = g + \nabla \cdot \underline{\tau}. \tag{1.62}$$

1.7   Let $\Sigma(t)$ be a family of surfaces depending on the parameter t (time). Define $\underline{v}_\Sigma(t)$ as the velocity of the points on the intersection of such surfaces with the orthogonal trajectories to them. It can be seen that $\underline{v}_\Sigma(t)$ fulfills

$$\underline{v}_\Sigma = v_\Sigma \underline{n}. \tag{1.63}$$

Here $\underline{n}$ is a unit normal vector and v_Σ is a scalar. Assume that for each t, the equation of the surface is given by

$$F(\underline{x}, t) = 0. \tag{1.64}$$

Show that

$$v_\Sigma = - \left(\frac{\partial F}{\partial x_i} \frac{\partial F}{\partial x_i} \right)^{-\frac{1}{2}} \frac{\partial F}{\partial t}. \tag{1.65}$$

1.8   Assume that a porous medium is saturated by a fluid that conserves its mass. Adopt the notations $\varepsilon(\underline{x}, t)$, $\rho(\underline{x}, t)$, and

$$\underline{u} \equiv \varepsilon \underline{v} \tag{1.66}$$

for the porosity, density, and Darcy velocity of the fluid, respectively. The fluid volume is an extensive property, since it is given by

$$E(t) = \int_{B(t)} \varepsilon(t) \, dx. \tag{1.67}$$

Prove that it satisfies the following global balance equation:

$$\frac{dE}{dt}(t) = - \int_{B(t)} \varepsilon \frac{\partial \operatorname{Ln} \rho}{\partial t} \, dx. \tag{1.68}$$

Therefore:

a) When the fluid is incompressible:

$$\frac{\partial \varepsilon}{\partial t} + \nabla \cdot (\underline{u}) = 0. \tag{1.69}$$

b) When the fluid is compressible:

$$\frac{\partial \varepsilon}{\partial t} + \nabla \cdot (\underline{u}) = -\varepsilon \frac{\partial \operatorname{Ln} \rho}{\partial t}. \tag{1.70}$$

1.9   Show that there is one-to-one correspondence between extensive and intensive properties.

1.10   Prove Theorem 5 of Section 1.4.2 with details.

1.11 Prove that for each body, the change of body volume per unit volume per unit time equals the divergence of the particle velocity; that is, $\nabla \cdot \underline{v}$. In particular, when there is *mass conservation*, show that

$$\nabla \cdot \underline{v} = \frac{D\rho^{-1}}{Dt}.$$

1.12 Prove Eq. (B-8).

REFERENCES

1. Allen, M.B., I. Herrera, and G.F. Pinder, *Numerical Modeling in Science and Engineering*, Wiley, New York, 1988.

2. Ciarlet, P.G., *The Finite Element Method for Elliptic Problems*, SIAM CL, 40, 1978.

3. Ciarlet, P.G. and J.L. Lions, *Handbook of Numerical Analysis,* Vol. II: Finite Element Methods (Part 1), Elsevier Science, Amsterdam, 1991.

4. Ciarlet, P.G. and J.L. Lions, *Handbook of Numerical Analysis*, Vol. III: Techniques of Scientific Computing (Part 1), Numerical Methods for Solids (Part 1), and Solution of Equations in R^n (Part 2), Elsevier Science, Amsterdam, 1994.

5. Ciarlet, P.G. and J.L. Lions, *Handbook of Numerical Analysis*, Vol. IV: Finite Element Methods (Part 2), Numerical Methods for Solids, Elsevier Science, Amsterdam, 1996.

6. Ciarlet, P.G., *Mathematical Elasticity*, Vol. II: Theory of Plates.North-Holland, Elsevier Amsterdam, 1997.

7. Ciarlet, P.G. and J.L. Lions, *Handbook of Numerical Analysis*, Vol. VI, Numerical Methods for Solids (Part 3), Numerical Methods for Fluids (Part 1), Elsevier Science, Amsterdam, 1998.

8. Eringen, A.C. and Craine, R.E., Continuum theories of mixtures: basic theory and historical development, Quart. *J. Mech. Appl. Math.*, 29(2), 209-244, 1976.

9. Eringen, A.C., *Mechanics of Continua*, 2nd ed., Krieger, Huntington, NY, 1980.

10. Gurtin, M.E., The linear theory of elasticity, *Handbuch der Physik*, Ed. S. Flügge, Vol. VIa/2, Springer-Verlag, Berlin, 1972.

11. Hellwig G., *Partial Differential Equations: An Introduction*, Blaisdell Publishing Co.(Ginn & Co.), New York, 1964.

12. Herrera, I. and A. Montalvo, *Modelos Matemáticos de Campos Geotérmicos Comunicaciones Técnicas*, IIMAS-UNAM, AN-295, 1982.

13. Herrera, I., A. Galindo, and R. Camacho, Shock modelling in variable bubble point problems of petroleum engineering, *Computational Modelling of Free and Moving Boundary Problems*, Vol. 1: Fluid Flow, Eds. L.C. Wrobel and C.A. Brebbia, CMP Books, San Francisco, CA, 399-415, 1991.

14. Herrera, I., R. Camacho, and A. Galindo, Shock modelling in petroleum engineering, Chapter 7 of *Computational Methods for Moving Boundary Problems in Heat and Fluid*

Flow, Eds. L.C. Wrobel and C.A. Brebbia, CMP Books, San Francisco, CA, 143-170, 1993.

15. Herrera, I., Shocks and bifurcations in black-oil models, *Soc. Petrol. Eng. J.*, 1(1), 51-58, 1996.

16. Herrera, I. and R.G. Camacho, A consistent approach to variable bubble-point systems, *Numer. Methods Partial Differential Equations*, 13(1), 1-18, 1997.

17. Hill, R., Discontinuity relations in mechanics of solids, *Progress in Solid Mechanics*. Vol. II(72), North-Holland, Amsterdam, 245-276, 1961.

18. Hunter, S.C., *Mechanics of Continuous Media*, Ellis Horwood, Chichester, UK, 1976.

19. Jeffrey, A., A note on the derivation of the discontinuity conditions across contact discontinuities, shocks and phase fronts. *Z. Angew. Math. Phys.*, 15, 68-71, 1964.

20. Keller, H.B., Propagation of stress discontinuities in inhomogeneous elastic media, *SIAM Rev.* 6, 356-382, 1964.

21. Lions, J.L. and E. Magenes, *Non-homogeneous Boundary Value Problems and Applications*, Springer-Verlag, Berlin (3 volumes), 1972.

22. Malvern, L. E., *Engineering Mechanics*, Prentice-Hall, Englewood Cliffs, NJ, 1976,.

23. Malvern, L.E. *Introduction to the Mechanics of a Continuous Medium,* Facsimile edition, Prentice-Hall, Englewood Cliffs, NJ, 1977.

24. Noll, W., A mathematical theory of the mechanical behavoir of continuous media, *Arch. Rational Mech. and Anal.,* 2, 197-226, 1958.

25. Noll, W., The Foundations of Mechanics and Thermodynamics (selected papers by W. Noll, with a preface by C. Truesdell), Springer-Verlag, Berlin, 1974.

26. Rutherford, A., *Mathematical Modelling Techniques*, Dover, New York, 1995.

27. Smirnov, V.I., *A Course of Higher Mathematics*, Pergamon Press, Oxford, UK,1964.

28. Sokolnikoff, I.S., *Mathematical Theory of Elasticity,* McGraw-Hill, New York, 1956.

29. Stoker, J.J., *Water Waves*, Interscience, New York, 1957.

30. Treves, F., *Basic Linear Partial Differential Equations*, Academic Press, New York, 1975.

31. Truesdell, C. and R. Toupin, The classical field theories, in Vols. 12, 15, 16, 17, 18, 19, 71, 72, and 73 of the *Handbuch der Physik*, Ed. S Flügge, Springer-Verlag, Berlin, 1960.

32. Truesdell, C., The rational mechanics of materials: past, present, future, *Appl. Mech. Rev.*, 12, 75-80, 1959. Reprinted in *Applied Mechanics Surveys,* Vol. 22, Spartan Books, Washington, DC, 1965.

33. Truesdell, C., General and exact theory of waves in finite elastic strain, *Arch. Rational Mech. Anal.,* 8, 263-296, 1961. Reprinted in B. Coleman, M.E. Gurtin, I. Herrera, and C. Truesdell, *Wave propagation in dissipative materials*, Springer-Verlag, Berlin, 1965.

34. Truesdell, C. and W. Noll, The non-linear field theories of mechanics, in Vols. 12, 16, 19, 20, 21, 23, and 73 of the *Handbuch der Physik*, Ed. S. Flügge, Springer-Verlag, Berlin, 1965.

35. Truesdell, C., *The Elements of Continuum Mechanics*, Vol 27, Springer-Verlag, Berlin, 1966.

36. Truesdell, C., A program toward rediscovering the rational mechanics of the age of reason, *Arch. Hist. Exact Sci.,* 1, 3-36, 1960. Corrected reprint in *Essays in the History of Mechanics,* Vol. 22, Springer-Verlag, Berlin, 1968.

37. Tychonov, A.N. and A.A. Samarski, *Partial Differential Equations of Mathematical Physics*, Holden-Day, San Francisco, CA, 1964.

38. Zheng, C.H. and G.D. Bennett, *Applied Contaminant Transport Modeling*, Van Nostrand Reinhold, Wiley, New York, 1995.

CHAPTER 2

MECHANICS OF CLASSICAL CONTINUOUS SYSTEMS

Taking advantage of the unified approach supplied by the axiomatic formulation of continuum mechanics, this chapter introduces the basic mathematical models of the mechanics of continuous systems which are fundamental to classical solid and fluid mechanics. In this chapter we only derive the basic mathematical models, leaving for later chapters detailed discussions of each of these topics.

2.1 ONE-PHASE SYSTEMS

The theory, that we have developed up to now applies to one-phase systems only. However, the classical solid and fluid systems to be treated in this chapter are one-phase systems, so they are suitable to illustrate the application of the theory thus far developed. A conspicuous feature of the theory of one-phase models of continuous systems is that at any time each point of the physical space is occupied by one, and only one, particle and, therefore, the particle velocity is uniquely defined at each point and time. However, more general models applicable to a wider class of systems encountered in science and engineering, for which several particle velocities may be defined at each point and time, will be introduced in Chapter 3.

Mathematical Modeling in Science and Engineering: An Axiomatic Approach.
By Ismael Herrera and George F. Pinder Copyright © 2012 John Wiley & Sons, Inc.

2.2 THE BASIC MATHEMATICAL MODEL OF ONE-PHASE SYSTEMS

The general procedure for constructing *one-phase mathematical models* is explained next. It can be briefly described as a two-step procedure:

1. A family of extensive properties is identified; and

2. The balance conditions are applied to each extensive property of the family.

The balance conditions are expressed by means of the differential equation of local balance and the jump condition. The resulting system of partial differential equations and jump conditions so obtained will be referred to as the *basic mathematical model of the system*.

So if N is the number of extensive properties of the system, then according to Eq. (1.20), each of them can be expressed as

$$E^\alpha (t) \equiv \int_{B(t)} \psi^\alpha(\underline{x}, t) \, d\underline{x}, \quad \alpha = 1, ..., N \tag{2.1}$$

where ψ^α is the corresponding intensive property. Furthermore, each of the extensive properties satisfies the global balance equation of Eq. (1.27):

$$\frac{dE^\alpha}{dt}(t) = \int_{B(t)} g^\alpha(\underline{x}, t) \, d\underline{x} + \int_{\partial B(t)} \underline{\tau}^\alpha(\underline{x}, t) \cdot \underline{n}(\underline{x}, t) \, d\underline{x} + \int_{\Sigma(t)} g_\Sigma^\alpha(\underline{x}, t) \, d\underline{x}. \tag{2.2}$$

Therefore, the mathematical model of the system is obtained by application of Eqs. (1.32) and (1.33) to each of the extensive properties. This yields the differential equations

$$\frac{\partial \psi^\alpha}{\partial t} + \boldsymbol{\nabla} \cdot (\underline{v}\psi^\alpha) = \boldsymbol{\nabla} \cdot \underline{\tau}^\alpha + g^\alpha, \quad \forall \alpha = 1, ..., N \tag{2.3}$$

and the jump conditions

$$[[\psi^\alpha (\underline{v}^\alpha - \underline{v}_\Sigma) - \underline{\tau}^\alpha]] \cdot \underline{n} = g_\Sigma^\alpha, \quad \forall \alpha = 1, ..., N. \tag{2.4}$$

These systems of differential equations and jump conditions constitute the above-defined basic mathematical model of the continuous system.

Generally, engineers and scientists are interested in developing *complete models*, that is, models capable of predicting system behavior when certain initial and boundary conditions are known. In order to obtain such models it is necessary to have adequate information about the system that is being modeled; this includes scientific and technological knowledge, as well as laboratory and field measurements, about the properties of the system. In particular, formulas that are called *constitutive equations*, which are established by means of scientific and technological theories and/or experiments, and permit expressing $g^\alpha(\underline{x}, t), g_\Sigma^\alpha(\underline{x}, t)$, and $\underline{\tau}^\alpha(\underline{x}, t)$ in terms of the intensive properties occurring in the model are needed; through them the scientific and technological knowledge about the system is incorporated into the model. In some cases, as in some solute-transport models, information about the particle velocities $\underline{v}^{\alpha}\left(\underline{x}, t\right)$ is also required in order to transform the basic mathematical model into a complete model.

A typical example of a constitutive equation is the *stress-strain relationship for elastic bodies*, but constitutive relationships may have very different forms. For example, if chemical phenomena occur in the system, the *chemical equilibrium or chemical kinetic equations* could be used to obtain the constitutive equations for g^{α}. In other models, such as some used in remediation strategies that are applied to subsurface hydrological systems, it is biological knowledge that is needed in order to obtain the constitutive equations for g^{α}. In summary, in addition to the two steps mentioned above that are used for constructing the basic mathematical model, there is another step that is required for transforming it into a complete model, capable of predicting the behavior of the system. It is necessary to obtain sufficient information about the system through laboratory or field measurements as well as constitutive equations that express the external supplies, g^{α}, and the fluxes, $\underline{\tau}^{\alpha}$, in terms of the other intensive properties involved in the model at hand. As noted earlier, in some models, such as the transport models to be discussed in Chapters 4 and 6, this information includes the particle velocities themselves.

The above-described complementary information about the system plays a very important role since it is the means by which scientific and technological knowledge about the system is incorporated into the models; thereby, such knowledge acquires the capacity of predicting behavior. Acquaintance with the theory of mathematical and computational modeling is needed for effectively and efficiently planning and performing experiments, since the mathematical model itself is a framework that clearly indicates the constitutive knowledge required to have a *complete model*.

Finally, after the steps mentioned above have been executed, a complete model that consists of a system of partial differential equations is generated. In order to predict system behavior it is necessary to formulate *well-posed problems* –that is, problems that possess one and only one solution for – such a system. Generally, such problems include initial and boundary conditions. In Appendix A, some basic concepts of partial differential equations, including a description of the most important well-posed problems for different types, is presented.

2.3 THE EXTENSIVE/INTENSIVE PROPERTIES OF CLASSICAL MECHANICS

The theory of solid and fluid mechanics as applied to single-phase systems may appropriately be called *the classical mechanics of continuous media*, because it was the first to be developed. Its implementation in mathematical modeling, as here presented, is based on the following family of extensive properties:

1. Mass;

2. Linear momentum;

3. Angular momentum;

4. Kinetic energy; and

5. Internal energy.

In this family, mass and energy are scalar-extensive properties, while linear and angular momentum are vector extensive properties (six properties). We observe that each vector-extensive property is equivalent to three scalar-extensive properties, and therefore when the balance conditions are expressed in terms of the intensive properties, a system of nine differential equations and nine jump conditions is obtained; this constitutes the basic mathematical model of the classical continuous system.

2.4 MASS CONSERVATION

A basic assumption of the mechanics of solids and fluids is the *mass conservation principle*; it states that no mass is created ($g \equiv 0$) and that no mass-diffusive processes occur ($\tau \equiv 0$), since the mass of any body is

$$M(t) = \int_{B(t)} \rho(\underline{x}, t) \, d\underline{x}. \tag{2.5}$$

Applying Eq. (2.2), it is seen that

$$\frac{dM}{dt}(t) = 0. \tag{2.6}$$

The *intensive property* associated with mass is the *density*. So mass conservation implies the differential equation [setting ψ^α to ρ in Eqs. (2.3) and (2.4)]

$$\frac{\partial \rho}{\partial t} + \boldsymbol{\nabla} \cdot (\rho \underline{v}) = 0 \tag{2.7}$$

and the jump condition

$$[\![\rho(\underline{v} - \underline{v}_\Sigma)]\!] \cdot \underline{n} = 0. \tag{2.8}$$

Equation (2.7) is also known as the *continuity equation*. Perhaps less recognized for its importance, Eq. (2.8) has a number of applications; a notable one is the study of shocks that occur in the supersonic flow of compressible fluids.

Notice that Eq. (2.8) can also be written as

$$\{\rho(\underline{v} - \underline{v}_\Sigma)\}_+ \cdot \underline{n} = \{\rho(\underline{v} - \underline{v}_\Sigma)\}_- \cdot \underline{n} \tag{2.9}$$

where the velocity of the fluid particles relative to the shock is given by $\underline{v} - \underline{v}_\Sigma$; thus, the expression $\rho(\underline{v} - \underline{v}_\Sigma) \cdot \underline{n}$ represents the *fluid mass that passes through the shock*. More specifically, Eq. (2.9) expresses the fact that the mass that enters the shock from one side is the same as the mass that goes out on the other side.

2.5 LINEAR MOMENTUM BALANCE

In *particle mechanics* the linear momentum of a particle is defined as $m\underline{v}$, where m and $\underline{v}$ are the *particle mass and velocity*, respectively. In the mechanics of continuous media, the *linear momentum of a body* is defined to be

$$\underline{M}(t) \equiv \int_{B(t)} \rho\,(\underline{x}, t)\,\underline{v}\,(\underline{x}, t)\, d\underline{x}. \tag{2.10}$$

Notice that the linear momentum is a vector-extensive property (since $\underline{v}$ is a vector quantity and, as such, has three scalar components). In particular, Eq. (2.10) is equivalent to three scalar equations: namely

$$\begin{aligned}
M_1\,(t) &\equiv \int_{B(t)} \rho\,(\underline{x}, t)\, v_1\,(\underline{x}, t)\, d\underline{x} \\
M_2\,(t) &\equiv \int_{B(t)} \rho\,(\underline{x}, t)\, v_2\,(\underline{x}, t)\, d\underline{x} \\
M_3\,(t) &\equiv \int_{B(t)} \rho\,(\underline{x}, t)\, v_3\,(\underline{x}, t)\, d\underline{x}.
\end{aligned} \tag{2.11}$$

Applying the general equation of global balance, Eq. (2.2), to each of the scalar extensive properties of Eq. (2.11), one obtains

$$\frac{dM_i}{dt}(t) = \int_{B(t)} g_i\,(\underline{x}, t)\, d\underline{x} + \int_{\partial B(t)} \underline{\tau}_i\,(\underline{x}, t) \cdot \underline{n}\,(\underline{x}, t)\, d\underline{x} + \int_{\Sigma(t)} g_\Sigma\,(\underline{x}, t)\, d\underline{x};$$
$$i = 1, 2, 3. \tag{2.12}$$

Notice that the vector $\underline{g}_\Sigma$, whose components are $g_{\Sigma i}$, is different from zero when there is a concentrated force on the surface Σ with intensity per unit area of $\underline{g}_\Sigma$. Let us define a new vector $\underline{b}$ as

$$\underline{b} \equiv \begin{pmatrix} b_1 \\ b_2 \\ b_3 \end{pmatrix} \equiv \rho^{-1} \underline{g} \tag{2.13}$$

so that $\underline{g} = \rho\underline{b}$. The vector $\underline{b}$ is usually called the *body force* and it represents the force per unit mass that is exerted by forces acting from a distance, such as those due to gravitational or electrical fields; it has units of acceleration (force divided by mass). We adopt the notation

$$\underline{\tau}_i \equiv (\tau_{i1}, \tau_{i2}, \tau_{i3}). \tag{2.14}$$

The external supply of linear momentum, $\underline{g}'$ has the physical interpretation of body force per unit volume. Then we define the *stress tensor* by

$$\underline{\underline{\sigma}} \equiv \begin{pmatrix} \sigma_{11}\sigma_{12}\sigma_{13} \\ \sigma_{21}\sigma_{22}\sigma_{23} \\ \sigma_{31}\sigma_{32}\sigma_{33} \end{pmatrix} \equiv \begin{pmatrix} \tau_{11}\tau_{12}\tau_{13} \\ \tau_{21}\tau_{22}\tau_{23} \\ \tau_{31}\tau_{32}\tau_{33} \end{pmatrix}. \tag{2.15}$$

In the case of the gravity force, for example, when the body is located on the Earth's surface, $\underline{b} = \hat{g}$, where $\hat{g}$ is the acceleration due to gravity. The *traction* on the surface of a body, whose unit outer normal vector is $\underline{n}$, is defined to be $\underline{T} \equiv \underline{\sigma} \cdot \underline{n}$ and it represents the force, per unit area, exerted on the body through the body surface by other bodies that are in contact with it. Observe that with this definition of the traction, such forces are directed toward the body exterior when $\underline{T} \cdot \underline{n} > 0$, a fact that agrees well with the common use of the word *traction*.

Above, we have decomposed the forces that act on a body into two general classes: forces that act at a distance, the body forces, and contact forces, the tractions. As for the global momentum balance of Eq. (2.12), in the absence of the concentrated forces, it can now be written as

$$\frac{d\underline{M}}{dt}(t) = \int_{B(t)} \rho\underline{b}\,(\underline{x}, t)\, d\underline{x} + \int_{\partial B(t)} \underline{\sigma}\,(\underline{x}, t) \cdot \underline{n}\,(\underline{x}, t)\, d\underline{x} \qquad (2.16)$$

or

$$\frac{d\underline{M}}{dt}(t) = \int_{B(t)} \rho\underline{b}\,(\underline{x}, t)\, d\underline{x} + \int_{\partial B(t)} \underline{T}\,(\underline{x}, t)\, d\underline{x}.$$

The first integral in the latter equation, which is over the volume of the body $B\,(t)$, represents the total force acting on the body (the body force) attributable to a force which has a distant origin, while the second integral defined over the surface $\partial B\,(t)$ represents the total *contact force*, a force due to the action of neighboring bodies. So this balance equation establishes that, just as in particle mechanics, the rate of change of linear momentum equals the total force acting on the body.

From Eq. (2.10) it follows that the intensive property associated with the linear momentum of a body is the product of its mass density and its particle velocity: $\rho\,(\underline{x}, t)\,\underline{v}\,(\underline{x}, t)$; it has three scalar components:

$$\rho\underline{v} = \begin{pmatrix} \rho v_1 \\ \rho v_2 \\ \rho v_3 \end{pmatrix}. \qquad (2.17)$$

We can now introduce Einstein's summation convention, which states that repeated indices are summed over their ranges (explained further in Appendix E). Hence, the application of Eqs. (1.32) and (1.33) to each of these scalar components yields (assuming that no concentrated body forces act on the body; that is, $\underline{g}_\Sigma = 0$)

$$\left. \begin{array}{l} \dfrac{\partial \rho v_i}{\partial t} + \nabla \cdot (\underline{v}\rho v_i) = \nabla \cdot \underline{\tau}_i + g_i \\[2mm] [\![\rho v_i\,(\underline{v} - \underline{v}_\Sigma) - \underline{\tau}_i]\!] \cdot \underline{n} = 0 \ \text{ on } \Sigma \end{array} \right\} \quad i = 1, 2, 3 \qquad (2.18)$$

or

$$\left. \begin{array}{l} \dfrac{\partial \rho v_i}{\partial t} + \dfrac{\partial \rho v_i v_j}{\partial x_j} = \dfrac{\partial \sigma_{ij}}{\partial x_j} + \rho b_i \\[2mm] [\![\rho v_i\,(v_j - v_{\Sigma j}) - \sigma_{ij}]\!] \cdot n_j = 0 \ \text{ on } \Sigma \end{array} \right\} \quad i = 1, 2, 3. \qquad (2.19)$$

Furthermore, when the conservation of mass conditions of Eqs. (2.7) and (2.8) are combined with Eq. (2.19), one obtains

$$\left.\begin{array}{c} \rho \left(\dfrac{\partial v_i}{\partial t} + v_j \dfrac{\partial v_i}{\partial x_j} \right) = \dfrac{\partial \sigma_{ij}}{\partial x_j} + \rho b_i \\[2mm] [\![v_i]\!]\, \rho\, (v_j - v_{\Sigma j})\, n_j - [\![\sigma_{ij}]\!]\, n_j = 0 \ \text{on} \ \Sigma \end{array}\right\} \quad i = 1,2,3. \qquad (2.20)$$

Observe that the meaning of Eq. (2.20) is unambiguous since Eq. (2.8) states that the product $\rho\,(\underline{v} - \underline{v}_\Sigma) \cdot \underline{n}$ is continuous across the discontinuity surface Σ; that is, its value on Σ is uniquely defined. The relations presented in Eq. (2.20) are frequently written as the following pair of equations:

$$\rho \left(\frac{\partial \underline{v}}{\partial t} + \underline{v} \cdot \nabla \underline{v} \right) = \nabla \cdot \underline{\underline{\sigma}} + \rho \underline{b} \qquad (2.21)$$

and

$$[\![\underline{v}]\!]\, \rho\, (\underline{v} - \underline{v}_\Sigma) \cdot \underline{n} - [\![\underline{\underline{\sigma}}]\!] \cdot \underline{n} = 0 \ \text{on} \ \Sigma. \qquad (2.22)$$

As noted earlier, there is no ambiguity in Eq. (2.22) because $\rho\,(\underline{v} - \underline{v}_\Sigma) \cdot \underline{n}$ is continuous across Σ. In Eq. (2.21) another possible source of ambiguity is the product $\underline{v} \cdot \nabla \underline{v}$; taking Eq. (2.20) into account it is seen that it must be understood as $(\underline{v} \cdot \nabla)\, \underline{v}$. Using the *material derivative*, defined as

$$\frac{D(\cdot)}{Dt} \equiv \frac{\partial\,(\cdot)}{\partial t} + \underline{v} \cdot \nabla\,(\cdot)$$

an alternative expression that is frequently used for Eq. (2.21) is obtained:

$$\rho \frac{D\underline{v}}{Dt} = \nabla \cdot \underline{\underline{\sigma}} + \rho \underline{b}. \qquad (2.23)$$

Taking the scalar product of this equation with $\underline{v}$, one obtains

$$\rho \frac{D \frac{1}{2} v^2}{Dt} = \underline{v} \cdot \nabla \cdot \underline{\underline{\sigma}} + \rho \underline{v} \cdot \underline{b}. \qquad (2.24)$$

This equation will be used later.

2.6   ANGULAR MOMENTUM BALANCE

In particle mechanics the *angular momentum* of a particle is defined to be

$$\underline{M}_a\,(t) \equiv \underline{x} \times m\underline{v}. \qquad (2.25)$$

where m is the mass particle and $\times$ is the cross product of two vectors. Motivated by this definition, in continuum mechanics the angular momentum is defined so that the angular momentum per unit volume of a body is the mass per unit volume times the vector product of the position vector by the particle velocity: that is, $\underline{x} \times \rho \underline{v}$. Thus, in continuum mechanics the definition of angular momentum of a body is

$$\underline{M}_a(t) = \int_{B(t)} \rho\left(\underline{x} \times \underline{v}\right) d\underline{x}. \tag{2.26}$$

Through the choice of this definition, we observe that in the limit when the mass of the body is concentrated at a single point, this expression reduces to that of particle mechanics, given by Eq. (2.25); therefore, the concept used in continuum mechanics is a generalization of that of particle mechanics. Clearly, in view of Eq. (2.26), the intensive property corresponding to the angular momentum is

$$\underline{\psi} \equiv \rho\left(\underline{x} \times \underline{v}\right). \tag{2.27}$$

On the other hand, the rate of change of angular momentum that a body experiences equals the total *torque* acting on the body, where the torque is defined as the vector product of a force and the length of a lever arm originating at $r = 0$. The total torque can be decomposed into two parts: the *torque due to body forces* and that *due to the tractions acting on the body*. The torque per unit volume at any point a of the interior of a body, due to body forces, equals the vector product of the force per unit volume and the position vector of the point considered; while the torque per unit area a at any point on the boundary of a body, due to tractions acting there, equals the vector product of the force per unit area acting on the body (that is, the traction $\underline{\underline{\sigma}} \cdot \underline{n}$) and the position vector of the point considered. This yields

$$\frac{d\underline{M}_a}{dt}(t) = \int_{B(t)} \rho\left(\underline{x} \times \underline{b}\right) d\underline{x} + \int_{\partial B(t)} \underline{x} \times \underline{\underline{\sigma}} \cdot \underline{n}\, d\underline{x} \tag{2.28}$$

for the global balance of angular momentum of a body. This equation reduces to Eq. (2.2) if we set $g^\alpha \equiv \rho\left(\underline{x} \times \underline{b}\right)$ and $\underline{\tau}^\alpha \equiv \underline{x} \times \underline{\underline{\sigma}}$.

Application of the differential equation of local balance, Eq. (2.3), yields

$$\frac{D\rho\left(\underline{x} \times \underline{v}\right)}{Dt} + \rho\left(\underline{x} \times \underline{v}\right) \nabla \cdot \underline{v} = \nabla \cdot \left(\underline{x} \times \underline{\underline{\sigma}}\right) + \rho\left(\underline{x} \times \underline{b}\right), \quad \forall x \in B(t). \tag{2.29}$$

Through expansion of the first term in Eq. (2.29), we obtain

$$\frac{D\rho\left(\underline{x} \times \underline{v}\right)}{Dt} = \left(\underline{x} \times \underline{v}\right)\frac{D\rho}{Dt} + \rho\frac{D\left(\underline{x} \times \underline{v}\right)}{Dt}. \tag{2.30}$$

If one subtracts the mass conservation equation from Eq. (2.30), one obtains

$$\frac{D\rho\left(\underline{x} \times \underline{v}\right)}{Dt} = \left(\underline{x} \times \underline{v}\right)\frac{D\rho}{Dt} - \left(\underline{x} \times \underline{v}\right)\left[\frac{\partial \rho}{\partial t} + \nabla\left(\rho\underline{v}\right)\right] + \rho\frac{D\left(\underline{x} \times \underline{v}\right)}{Dt}$$

which upon expansion of $\left(\underline{x} \times \underline{v}\right)\frac{D\rho}{Dt}$ yields

$$\frac{D\rho\left(\underline{x} \times \underline{v}\right)}{Dt} + \rho\left(\underline{x} \times \underline{v}\right) \nabla \cdot \underline{v} = \rho\frac{D\left(\underline{x} \times \underline{v}\right)}{Dt}. \tag{2.31}$$

Also, one can show via the definition of the cross derivative that

$$\frac{D\left(\underline{x} \times \underline{v}\right)}{Dt} = \underline{x} \times \frac{D\underline{v}}{Dt} \tag{2.32}$$

so that using

$$\rho \frac{D\underline{v}}{Dt} = \boldsymbol{\nabla} \cdot \underline{\underline{\sigma}} + \rho \underline{b}$$

we obtain

$$\rho \underline{x} \times \frac{D\underline{v}}{Dt} = \boldsymbol{\nabla} \cdot \left(\underline{x} \times \underline{\underline{\sigma}}\right) + \rho \left(\underline{x} \times \underline{b}\right). \tag{2.33}$$

However,

$$\boldsymbol{\nabla} \cdot \left(\underline{x} \times \underline{\underline{\sigma}}\right) = \underline{x} \times \left(\boldsymbol{\nabla} \cdot \underline{\underline{\sigma}}\right) + \left(\boldsymbol{\nabla}\underline{x}\right) \times \underline{\underline{\sigma}}. \tag{2.34}$$

Thus substitution of Eq. (2.34) into Eq. (2.33) gives

$$\rho \underline{x} \times \frac{D\underline{v}}{Dt} = \underline{x} \times \left(\boldsymbol{\nabla} \cdot \underline{\underline{\sigma}}\right) + \left(\boldsymbol{\nabla}\underline{x}\right) \times \underline{\underline{\sigma}} + \rho \underline{x} \times \underline{b}. \tag{2.35}$$

Taking $\underline{x} \times$ as a common factor, we obtain

$$\underline{x} \times \left\{ \rho \frac{D\underline{v}}{Dt} - \boldsymbol{\nabla} \cdot \underline{\underline{\sigma}} - \rho \underline{b} \right\} - \left(\boldsymbol{\nabla}\underline{x}\right) \times \underline{\underline{\sigma}} = 0. \tag{2.36}$$

Using the fact that $\rho \frac{D\underline{v}}{Dt} - \boldsymbol{\nabla} \cdot \underline{\underline{\sigma}} - \rho \underline{b} = 0$, this reduces to

$$\left(\boldsymbol{\nabla}\underline{x}\right) \times \underline{\underline{\sigma}} = 0. \tag{2.37}$$

The elements of the matrix $\boldsymbol{\nabla}\underline{x}$ are $\partial x_i / \partial x_j = \delta_{ij}$; which is equivalent to saying that $\boldsymbol{\nabla}\underline{x} = \underline{\underline{I}}$, where $\underline{\underline{I}}$ is the identity matrix. Expressing Eq. (2.37) by means of the Levi Civita symbol this equation becomes

$$\sum_{j,k,l=1}^{3} \in_{ijk} \delta_{jl}\sigma_{lk} = \sum_{j,k,l=1}^{3} \in_{ijk}\sigma_{jk} = 0; \quad i = 1, 2, 3. \tag{2.38}$$

This equation is a vector equation that can be expressed as

$$\left(\sigma_{23} - \sigma_{32}, \sigma_{31} - \sigma_{13}, \sigma_{12} - \sigma_{21}\right) = 0; \tag{2.39}$$

that is,

$$\sigma_{23} - \sigma_{32} = 0; \quad \sigma_{31} - \sigma_{13} = 0; \quad \sigma_{12} - \sigma_{21} = 0. \tag{2.40}$$

In summary, the stress tensor is symmetric; that is,

$$\underline{\underline{\sigma}} = \underline{\underline{\sigma}}^T. \tag{2.41}$$

To benefit from the simplicity of this result, in what follows it will always be understood that the stress tensor is symmetric, which grants that the angular momentum balance is fulfilled.

2.7   ENERGY CONCEPTS

The present-day understanding of heat and its relation to work was developed during the second half of the nineteenth century. Crucial to this understanding were the experiments of James Prescott Joule [37]. Experiments such as those conducted by Joule led to the concept that bodies contain energy in several forms, one of them being *mechanical energy*, but can also store it in another form, called *internal energy*. The recognition of heat and internal energy as forms of energy makes it possible to formulate the *First Law of Thermodynamics*:\Although energy assumes many forms, the total quantity of energy is constant, and when energy disappears in one form it reappears in other form". All of these concepts are incorporated in the following development.

Let $B\,(t)$ be a body, then its *kinetic energy* (also called *mechanical energy*), $E_K\,(t)$, and its internal energy, $E_I\,(t)$, are defined to be extensive properties. The former is given by

$$E_K\,(t) \equiv \int_{B(t)} \tfrac{1}{2}\rho v^2 \, d\underline{x}. \tag{2.42}$$

The notation $U\,(\underline{x}, t)$ is adopted to describe the internal energy per unit mass of such a body, so that the intensive property associated with the internal energy, as an extensive property, is $\rho U\,(\underline{x}, t)$. Therefore,

$$E_I\,(t) \equiv \int_{B(t)} \rho U \, d\underline{x}. \tag{2.43}$$

It will be useful to have available the concept of *total energy*, or simply, *energy*. It is defined by

$$E\,(t) = E_I\,(t) + E_K\,(t). \tag{2.44}$$

Clearly, in view of Eqs. (2.42) to (2.44), one has

$$E\,(t) \equiv \int_{B(t)} \rho \left(U + \tfrac{1}{2}v^2\right) d\underline{x}. \tag{2.45}$$

We will return to this relationship shortly.

The energy balance equations that we are about to derive are based on a group of *mechanical axioms* in which the concepts discussed above are incorporated: such mechanical axioms are:

1. Kinetic energy and internal energy can be transformed from one to the other. Furthermore, the sum of the internal energy that is transformed into kinetic energy plus the kinetic energy that is transformed into internal energy vanishes;

2. The change in the kinetic energy of a body equals the work done by the forces acting on it, plus the internal energy that is transformed into kinetic energy (sometimes loosely called mechanical energy); and

3. The change in the internal energy of a body equals the internal energy generated by sources distributed in the interior of the body, plus the heat that flows into

the body through its boundary, plus the kinetic energy that is transformed into internal energy.

2.8 THE BALANCE OF KINETIC ENERGY

These principles imply that for the kinetic energy we have

$$\frac{dE_K}{dt}(t) = \int_{B(t)} \left(\rho \underline{b} \cdot \underline{v} + g_I^K \right) d\underline{x} + \int_{\partial B(t)} \underline{v} \cdot \underline{\underline{\sigma}} \cdot \underline{n} \, d\underline{x} + \int_{\Sigma(t)} g_{\Sigma I}^K \, d\underline{x}. \quad (2.46)$$

Here, $\rho \underline{b} \cdot \underline{v}$ is the rate (per unit volume) at which the body forces do work on the body; recall that $\rho \underline{b}$ represents the body forces per unit volume acting in the interior of the body. Furthermore, the term

$$\int_{\partial B(t)} \underline{v} \cdot \left(\underline{\underline{\sigma}} \cdot \underline{n} \right) d\underline{x} = \int_{\partial B(t)} \underline{v} \cdot \underline{T} \cdot \underline{n} \, d\underline{x} \quad (2.47)$$

is the rate (per unit area) at which the tractions, $\underline{T}$, acting on the external boundary of the body do work. The term g_I^K is the rate (per unit volume) at which internal energy is transformed into kinetic energy, which in view of Joule's principle is generally different from zero. Furthermore, $g_{\Sigma I}^K$ is the rate (per unit area) at which internal energy is transformed into kinetic energy at the shock Σ.

Equation (2.46) becomes the equation of global balance, Eq. (2.2), if we set

$$g \equiv \rho \underline{b} \cdot \underline{v} + g_I^K, \quad g_\Sigma \equiv g_{\Sigma I}^K, \quad \text{and} \quad \underline{\tau} \equiv \underline{\underline{\sigma}} \cdot \underline{v}. \quad (2.48)$$

Therefore, in view of the local balance conditions, Eqs. (2.3) and (2.4), one has

$$\frac{\partial \left(\frac{1}{2} \rho v^2 \right)}{\partial t} + \frac{1}{2} \nabla \cdot \left\{ \rho v^2 \underline{v} \right\} = \nabla \cdot \left(\underline{\underline{\sigma}} \cdot \underline{v} \right) + \rho \underline{b} \cdot \underline{v} + g_I^K; \quad \forall \underline{x} \in B(t) \quad (2.49)$$

together with

$$\left[\!\left[\frac{1}{2} \rho v^2 \left(\underline{v} - \underline{v}_\Sigma \right) - \underline{\underline{\sigma}} \cdot \underline{v} \right]\!\right] \cdot \underline{n} = g_{\Sigma I}^K; \quad \forall \underline{x} \in \Sigma. \quad (2.50)$$

Combination of Eq. (2.49) with the continuity equation yields

$$\rho \frac{D \left(\frac{1}{2} v^2 \right)}{Dt} = \underline{v} \cdot \nabla \cdot \underline{\underline{\sigma}} + \underline{\underline{\sigma}} : \nabla \underline{v} + \rho \underline{b} \cdot \underline{v} + g_I^K \quad (2.51)$$

where the : notation represents the scalar product of tensor quantities. Equation (2.51) in turn reduces to

$$\underline{\underline{\sigma}} : \nabla \underline{v} + g_I^K = 0 \quad (2.52)$$

by virtue of Eq. (2.24). Equation (2.52) is especially relevant, as soon will be seen.

2.9   THE BALANCE OF INTERNAL ENERGY

As for the internal energy, we have

$$\frac{dE_I}{dt}(t) = \int_{B(t)} \left(\rho h + g_K^I\right) d\underline{x} + \int_{\partial B(t)} \underline{q} \cdot \underline{n}\, d\underline{x} + \int_{\Sigma(t)} g_{\Sigma I}^K\, d\underline{x} \qquad (2.53)$$

where h represents *the rate (per unit mass; so that ρh is the rate per unit volume) at which internal energy is supplied by sources distributed in the body interior*; such sources are due, for example, to chemical reactions. As for $\underline{q} \cdot \underline{n}$, *it is the rate (per unit area) at which heat enters the external boundary of the body*, which is usually called the *heat flux*. The terms g_K^I and $g_{\Sigma I}^K$ *are the rates (per unit volume and per unit area, respectively) at which kinetic energy is transformed into internal energy.* According to Joule's principle, $g_K^I + g_I^K = 0$ and $g_{\Sigma K}^I + g_{\Sigma I}^K = 0$. Equation (2.53) becomes the equation of global balance, Eq. (2.2), if we set

$$g \equiv \rho h + g_K^I, \ \ g_\Sigma \equiv g_{\Sigma K}^I, \ \ \text{and} \ \ \underline{\tau} \equiv \underline{q}. \qquad (2.54)$$

Therefore, in view of the local balance equations, Eqs. (2.3) and (2.4), one has

$$\frac{\partial \rho U}{\partial t} + \boldsymbol{\nabla} \cdot (\rho U \underline{v}) = \boldsymbol{\nabla} \cdot \underline{q} + \rho h + g_K^I \qquad (2.55)$$

together with

$$\left[\!\left[\rho U \left(\underline{v} - \underline{v}_\Sigma\right) - \underline{q} \right]\!\right] \cdot \underline{n}_\Sigma = g_{\Sigma K}^I. \qquad (2.56)$$

Combining Eq. (2.55) with the continuity equation, one obtains

$$\rho \frac{DU}{Dt} = \boldsymbol{\nabla} \cdot \underline{q} + \rho h + g_K^I. \qquad (2.57)$$

Similarly, Eq. (2.50) can be written as

$$\begin{aligned}
&\rho \dot{\overline{v}} \left(\underline{v} - \underline{v}_\Sigma\right) \cdot [\![v]\!] - \left[\!\left[\underline{\underline{\sigma}} \cdot \underline{v} \right]\!\right] \cdot \underline{n} \\
&= \ \dot{\overline{v}} \left\{ \rho \left(\underline{v} - \underline{v}_\Sigma\right) \cdot \underline{n}\, [\![v]\!] - \left[\!\left[\underline{\underline{\sigma}} \right]\!\right] \cdot \underline{n} \right\} - [\![v]\!] \cdot \dot{\underline{\underline{\sigma}}} \cdot \underline{n} \\
&= \ g_{\Sigma K}^I; \ \ \forall \underline{x} \in \Sigma.
\end{aligned} \qquad (2.58)$$

To obtain Eq. (2.58) we combined Eq. (2.50) with the mass jump conditions of Eq. (2.8). Using Eq. (2.22), it follows that

$$g_{\Sigma K}^I = -\dot{\underline{\underline{\sigma}}}\, [\![v]\!] \cdot \underline{n}; \ \ \forall \underline{x} \in \Sigma. \qquad (2.59)$$

On the other hand, the jump conditions of (2.56) can be written, in view of Eq. (2.8), as

$$[\![U]\!]\, \rho \left(\underline{v} - \underline{v}_\Sigma\right) \cdot \underline{n} - \left[\!\left[\underline{q} \right]\!\right] \cdot \underline{n} = g_{\Sigma K}^I; \ \ \forall \underline{x} \in \Sigma \qquad (2.60)$$

Adding this equation to Eq. (2.59) yields

$$[\![U]\!]\, \rho \left(\underline{v} - \underline{v}_\Sigma\right) \cdot \underline{n} - \left[\!\left[\underline{q} \right]\!\right] \cdot \underline{n} = \dot{\underline{\underline{\sigma}}}\, [\![v]\!] \cdot \underline{n} = 0. \qquad (2.61)$$

2.10 HEAT EQUIVALENT OF MECHANICAL WORK

For the total energy, adding Eqs. (2.46) and (2.53), one obtains

$$\frac{dE}{dt}(t) = \int_{B(t)} \left\{ \rho \left(\underline{b} \cdot \underline{v} + h \right) + g_I^K + g_K^I \right\} d\underline{x} + \int_{\partial B(t)} \left\{ \underline{v} \cdot \underline{T} + q \cdot \underline{n} \right\} d\underline{x}. \quad (2.62)$$

When the body is isolated, all the external actions vanish and this equation implies that

$$\int_{B(t)} \left\{ g_I^K + g_K^I \right\} d\underline{x} = \frac{dE}{dt}(t) = 0 \quad (2.63)$$

since

$$g_K^K + g_I^K = 0. \quad (2.64)$$

Now Eqs. (2.52) and (2.64), together, imply that

$$g_K^I = \underline{\underline{\sigma}} : \boldsymbol{\nabla} \underline{v}. \quad (2.65)$$

This equation supplies an explicit expression for the rate at which mechanical energy is transformed into internal energy. Using it, Eq. (2.57) becomes

$$\rho \frac{DU}{Dt} = \boldsymbol{\nabla} \cdot q + \rho h + \underline{\underline{\sigma}} : \boldsymbol{\nabla} \underline{v}. \quad (2.66)$$

In this last equation it is advantageous that the rates of exchange between *heat* and *kinetic energy*, g_K^I and g_I^K, which generally are not known beforehand, have been eliminated. Equation (2.66) is fundamental for modeling thermodynamical systems and heat transport.

2.11 SUMMARY OF BASIC EQUATIONS FOR SOLID AND FLUID MECHANICS

No.	Extensive Properties	Intensive Properties ψ	τ	g	Balance Equation
1	Mass $M(t)$	ρ	0	0	$\frac{D\rho}{Dt} + \rho \nabla \cdot \underline{v} = 0$
2	Linear Momentum $M(t)$	$\rho \underline{v}$	$\underline{\underline{\sigma}}$	$\rho \underline{b}$	$\rho \frac{D\underline{v}}{Dt} - \nabla \cdot \underline{\underline{\sigma}} - \rho \underline{b} = 0$
3	Angular Momentum $M_a(t)$	$\rho\,(\underline{x} \times \underline{v})$	$\underline{x} \times \underline{\underline{\sigma}}$	$\rho\,(\underline{x} \times \underline{b})$	$\underline{\underline{\sigma}} = \underline{\underline{\sigma}}^T$
4	Total Energy $E(t)$	$\rho\left(E + \frac{1}{2}|\underline{v}|^2\right)$	$q + \underline{\underline{\sigma}} \cdot \underline{v}$	$\rho\,(h + \underline{b} \cdot \underline{v})$	$\rho \frac{DE}{Dt} = \nabla \cdot q + \rho h + \underline{\underline{\sigma}} : \nabla \underline{v}$

2.12   SOME BASIC CONCEPTS OF THERMODYNAMICS

Let us begin by dividing Eq. (2.66) by ρ, whereby one obtains

$$\frac{DU}{Dt} = \frac{1}{\rho} \boldsymbol{\nabla} \cdot \underline{q} + h + \frac{1}{\rho} \underline{\underline{\sigma}} : \boldsymbol{\nabla} \underline{v}. \tag{2.67}$$

Many developments in *thermodynamics* have considered the special case of an isotropic stress-tensor. In this instance the tractions are always perpendicular to the surface where they occur, so that

$$\underline{\underline{\sigma}} = -p\underline{\underline{I}}. \tag{2.68}$$

The scalar function $p\left(\underline{x}, t\right)$ is the *pressure*. Equation (2.68) can also be written using indicial notation as

$$\sigma_{ij} = -p\delta_{ij}. \tag{2.69}$$

In view of Eqs. (2.65) and (2.69), the rate at which kinetic energy is transformed into internal energy can be expressed as follows:

$$\underline{\underline{\sigma}} : \boldsymbol{\nabla} \underline{v} = \sigma_{ij} \frac{\partial v_i}{\partial x_j} = -p\delta_{ij} \frac{\partial v_i}{\partial x_j} = -p \frac{\partial v_i}{\partial x_i} = -p\boldsymbol{\nabla} \cdot \underline{v}. \tag{2.70}$$

According to the equation of continuity, one has

$$\boldsymbol{\nabla} \cdot \underline{v} = -\frac{1}{\rho} \frac{D\rho}{Dt}. \tag{2.71}$$

Therefore, using this equation and Eq. (2.67), we obtain

$$\frac{DU}{Dt} = \frac{1}{\rho} \boldsymbol{\nabla} \cdot \underline{q} + h + \frac{p}{\rho^2} \frac{D\rho}{Dt}. \tag{2.72}$$

The volume per unit mass is called *specific volume*, $V \equiv \rho^{-1}$; introducing it in this equation, one gets

$$\frac{DU}{Dt} = \frac{1}{\rho} \boldsymbol{\nabla} \cdot \underline{q} + h - p \frac{DV}{Dt}. \tag{2.73}$$

The case when the process is *adiabatic*, in which no heat sources occur and thermal conduction vanishes, that is, when $\boldsymbol{\nabla} \cdot \underline{q} = 0$ and $h = 0$, has special interest. Then Eq. (2.73) becomes

$$\frac{DU}{Dt} + p \frac{DV}{Dt} = 0. \tag{2.74}$$

2.12.1   Heat transport

The starting point for the mathematical modeling of heat transport is Eq. (2.67). For clarity in presentation, throughout this section the stress tensor will be assumed to be isotropic. We begin by rewriting Eq. (2.73) with the specific volume replaced by ρ^{-1} as

$$\frac{DU}{Dt} + p\frac{D\rho^{-1}}{Dt} - \frac{1}{\rho}\boldsymbol{\nabla} \cdot \underline{q} = h. \tag{2.75}$$

For some thermodynamical processes, such as *isobaric processes* (constant pressure), the density and internal energy per unit mass are functions of the temperature exclusively: that is, $\rho(T)$ and $U(T)$. To simplify somewhat the following discussions of this section, it is specifically assumed that the processes are isobaric, although the assumption that is essential is that ρ and E are functions of the temperature exclusively. As an example, consider the equation of state of an ideal gas:

$$\frac{p}{\rho} = nRT. \tag{2.76}$$

Here, n is a constant that depends on the gas molecular weight. For constant pressure, one has

$$\rho(T) \equiv \frac{p}{nR}T^{-1}. \tag{2.77}$$

In a similar fashion, for isobaric processes of a wider class of materials, it is possible to establish that E is also a function of the temperature exclusively. Incorporating the functions $E(T(\underline{x},t))$ and $\rho(T(\underline{x},t))$ in Eq. (2.75), one obtains

$$\left(\frac{dU}{dT} + p\frac{d\rho^{-1}}{dT}\right)\frac{DT}{Dt} - \frac{1}{\rho}\boldsymbol{\nabla} \cdot \underline{q} = h \tag{2.78}$$

which can be written as

$$c_p\frac{DT}{Dt} - \frac{1}{\rho}\boldsymbol{\nabla} \cdot \underline{q} = h \tag{2.79}$$

or

$$c_p\left(\frac{\partial T}{\partial t} + \underline{v} \cdot \boldsymbol{\nabla} T\right) - \frac{1}{\rho}\boldsymbol{\nabla} \cdot \underline{q} = h$$

where c_p is the *specific heat at constant pressure*, defined by

$$c_p \equiv \frac{dU}{dT} + p\frac{d\rho^{-1}}{dT}. \tag{2.80}$$

To obtain a complete model for the temperature it is necessary to have a constitutive equation expressing the heat flux vector, $\underline{q}$, in terms of the temperature distribution; one that is widely used is *Fourier's law*. This law states that

$$\underline{q} = \underline{\underline{k}}_H \boldsymbol{\nabla} T. \tag{2.81}$$

Here $\underline{\underline{k}}_H$ is the matrix of *thermal conductivity*. Then

$$\rho c_p\left(\frac{\partial T}{\partial t} + \underline{v} \cdot \boldsymbol{\nabla} T\right) - \boldsymbol{\nabla} \cdot \left(\underline{\underline{k}}_H \boldsymbol{\nabla} T\right) = \rho h. \tag{2.82}$$

When the matrix of thermal conductivity is isotropic, $\underline{\underline{k}}_H = k_H \underline{\underline{I}}$. Then

$$\rho c_p \left(\frac{\partial T}{\partial t} + \underline{v} \cdot \nabla T \right) - \nabla \cdot (k_H \nabla T) = \rho h. \qquad (2.83)$$

If the material is homogeneous, then

$$\frac{\partial T}{\partial t} + \underline{v} \cdot \nabla T - \underline{\underline{\kappa}}_H : \nabla \cdot \nabla T = \frac{h}{c_p}. \qquad (2.84)$$

Here $\underline{\underline{\kappa}}_H = \frac{\underline{\underline{k}}_H}{\rho c_p}$ is the *thermal diffusivity matrix*. When it is, in addition, isotropic, $\underline{\underline{\kappa}}_H = \kappa_H \underline{\underline{I}}$, then

$$\frac{\partial T}{\partial t} + \underline{v} \cdot \nabla T - \kappa_H \Delta T = \frac{h}{c_p}. \qquad (2.85)$$

If the continuous system is at rest, $\underline{v} = 0$, the classical equation of heat diffusion is obtained:

$$\frac{\partial T}{\partial t} - \kappa_H \Delta T = \frac{h}{c_p}. \qquad (2.86)$$

At this point we have presented the classical mechanics of continuous media and turn our attention in the next chapter to a family of models that, although they are not as classical, are very relevant in many problems of engineering and science.

2.13 SUMMARY

In this chapter we have introduced the concept of the basic mathematical model for a one-phase continuous system and an axiomatic procedure for constructing it. Such a procedure can be applied to any one-phase system and, as an illustration of its applicability, we have developed the basic mathematical models of the systems of classical mechanics of continuous media. The models presented here constitute an essential piece of the theoretical foundations of continuum mechanics, and in our presentation they were based on the following extensive properties: mass, linear momentum, angular momentum and finally, internal and kinetic energy. A conspicuous feature of the methodology on which this book is based has permitted us to separate appropriately the analysis of internal and kinetic energy; more standard approaches put them together in a single extensive property: the total energy. It was shown that the angular momentum can be removed from the list of basic extensive properties since it was shown to be equivalent to the condition that the stress tensor be symmetric. The various equations arising from these sections are summarized in a convenient table. The chapter closes with a presentation of some thermodynamic concepts and the basic mathematical models of heat transport.

EXERCISES

There are a certain number of models for which the *physical space* is not three-dimensional. Students are usually familiar with some cases in which the models are two- or one-dimensional. In such cases, each *body*, $B(t)$, occupies a domain of the corresponding Euclidean space. A more general form of Eqs. (2.1) and (2.2), applicable to such cases, is

$$
\left.
\begin{aligned}
E^\alpha(t) &= \int_{B(t)} \psi^\alpha(\underline{x}, t)\, d\mu \\
\frac{dE^\alpha}{dt}(t) &= \int_{B(t)} g^\alpha(\underline{x}, t)\, d\mu + \int_{\partial B(t)} \underline{\tau}^\alpha(\underline{x}, t) \cdot \underline{n}\, d\mu \\
&\quad + \int_{\Sigma(t)} g_\Sigma^\alpha(\underline{x}, t)\, d\mu
\end{aligned}
\right\} ;
\qquad (2.87)
$$

$$
\alpha = 1, ..., N.
$$

Here, $d\mu$ is used to denote the element of the Euclidean measure of the corresponding space. If the physical space is two-dimensional, then $B(t)$ is a two-dimensional region (or domain), $\partial B(t)$ is a closed curve; and $\Sigma(t)$ is a curve (generally, not closed). Therefore,

$$
\int_{B(t)} g^\alpha(\underline{x}, t)\, d\mu, \quad \int_{\partial B(t)} \underline{\tau}^\alpha(\underline{x}, t)\, d\mu, \quad \text{and} \quad \int_{\Sigma(t)} g_\Sigma^\alpha(\underline{x}, t)\, d\mu
\qquad (2.88)
$$

are, respectively, an integral over $B(t)$, with respect to area; an integral over $\partial B(t)$, with respect to arc length; and an integral over $\Sigma(t)$, also with respect to arc length.

If the physical space is one-dimensional, then $B(t)$ is a one-dimensional domain and, for simplicity, we restrict attention to the case when it is an *open* segment whose endpoints are $a(t)$ and $b(t)$, with $a(t) < b(t)$; as for $\Sigma(t)$, it is a finite collection of points: $\{x_{\Sigma 1}, ..., x_{\Sigma M}\}$. Therefore,

$$
\int_{B(t)} g^\alpha(\underline{x}, t)\, d\mu, \quad \int_{\partial B(t)} \underline{\tau}^\alpha(\underline{x}, t) \cdot \underline{n}\, d\mu \quad \text{and} \quad \int_{\Sigma(t)} g_\Sigma^\alpha(\underline{x}, t)\, d\mu
\qquad (2.89)
$$

are, respectively,

$$
\int_{a(t)}^{b(t)} g^\alpha(\underline{x}, t)\, dx, \quad \int_{\partial B(t)} \underline{\tau}^\alpha(\underline{x}, t) \cdot \underline{n}\, d\mu = \tau^\alpha(b(t), t) - \tau^\alpha(a(t), t) \quad \text{and}
$$

$$
\sum_{x_{\Sigma i} \in B(t)} g_\Sigma^\alpha(x_{\Sigma i}, t).
\qquad (2.90)
$$

2.1   Prove that Eqs. (2.3) and (2.4) are also satisfied in two-dimensional problems.

2.2   Prove that in one-dimensional problems, Eqs. (2.3) and (2.4) have to be replaced by

$$
\frac{\partial \psi^\alpha}{\partial t} + \frac{\partial}{\partial x}(v \psi^\alpha) = \frac{\partial \tau^\alpha}{\partial x} + g^\alpha; \quad \alpha = 1, ..., N
\qquad (2.91)
$$

and

$$\llbracket \psi^\alpha \left(v^\alpha \left(x_\beta, t \right) - v_{\Sigma\beta} \right) \rrbracket = g_{\Sigma\beta}^\alpha \left(t \right) ; \quad \alpha = 1, ..., N, \quad \beta = 1, ..., M. \tag{2.92}$$

2.3   Define one extensive property for each of the following two cases: two- and one-dimensional physical spaces.

2.4   When a river is seen on a map, its course appears as a line, generally curved, but it can also be straight. For simplicity, let the course be straight. Consider exclusively the following processes that are occurring in the river: There is distributed runoff that enters the river though its banks and concentrated inflow of sewage that enters the river exclusively at a single point x_Σ. Observe that in this case any open segment $(a\left(t\right), b\left(t\right))$, where $a\left(t\right)$ and $b\left(t\right)$ move with the velocity of river water, characterizes a *body*. The mass of water contained in each such body defines an extensive property (i.e., a body function) given by

$$E(t) = \int_{a(t)}^{b(t)} \psi\left(x, t\right) dx. \tag{2.93}$$

Give a physical interpretation of the intensive property $\psi\left(x, t\right)$ associated with such an extensive property.

2.5   Establish the following balance equation for such an extensive property:

$$\frac{dE}{dt}(t) = \int_{a(t)}^{b(t)} g\left(x, t\right) dx + \begin{cases} g_\Sigma & \text{if } x_\Sigma \in \left(a, b\right) \\ 0 & \text{if } x_\Sigma \notin \left(a, b\right) \end{cases} \tag{2.94}$$

Give the physical interpretation of $g\left(x, t\right)$ and g_Σ^1. Compare this equation with Eq. (2.2) and discuss the similarities between them.

2.6   Show that for this model the local balance equations are

$$\frac{\partial \psi}{\partial t} + \frac{\partial \left(v\psi\right)}{\partial x} = g, \quad x \neq x_\Sigma \tag{2.95}$$

and

$$\llbracket \psi v \rrbracket = g_\Sigma \text{ at } x = x_\Sigma. \tag{2.96}$$

2.7   As defined in Section 1.4.4, a continuous system is said to be *isochoric* when each body conserves its volume. Furthermore, if the nature of a continuous system is such that it can only perform isochoric motions, then the system is said to be *incompressible*. Prove that when a continuous system is incompressible, the differential equation of mass conservation reduces to

$$\frac{\partial \rho}{\partial t} + \underline{v} \cdot \nabla \rho = 0. \tag{2.97}$$

This can be expressed, introducing the material derivative, as

$$\frac{D\rho}{Dt} = 0. \tag{2.98}$$

A fluid is said to be homogeneous when its properties are independent of position. Show that when the fluid is incompressible and homogeneous, mass conservation is tantamount to

$$\frac{\partial \rho}{\partial t} = 0. \tag{2.99}$$

2.8   The use of vector equations combined with the use of the symbol ∇ may lead to very compact expressions for a significant number of relations. However, every vector equation is equivalent to a system of three scalar equations, which when it is written in indicial notation can also be written as a single equation. When the indicial notation is combined with *Einstein's summation convention,* the simplicity so achieved is also remarkable. For example, Eq. (2.21) is a very compact vector equation, which when it is written in indicial notation attains the form

$$\rho \left(\frac{\partial v_i}{\partial t} + v_j \frac{\partial v_i}{\partial x_j} \right) = \frac{\partial \sigma_{ij}}{\partial x_j} + \rho b_i. \tag{2.100}$$

Furthermore, a disadvantage of the vector notation, with respect to indicial notation, is that very few people know the algebra of the symbol ∇, while almost everybody working in engineering and physical-mathematical sciences knows the algebra of derivatives of scalar functions. As an illustration, using indicial notation, show that the vector equation

$$\frac{\partial \rho \underline{v}}{\partial t} + \nabla \cdot (\rho \underline{v} \underline{v}) = \nabla \cdot \underline{\underline{\sigma}} + \rho \underline{b} \tag{2.101}$$

is equivalent to Eq. (2.100), when mass is conserved, i.e., when Eq. (2.24) is fulfilled.

2.9   Assuming mass conservation and the balance of linear momentum, Eq. (2.21), show that Eqs. (2.51) and (2.52) are equivalent.

2.10   Show that in the mechanics of classical continuous systems, at shocks, the rate at which internal energy is transformed into kinetic energy is given by

$$g_{\Sigma I}^K = -\underline{\dot{\sigma}} \left[\![\underline{v}]\!\right] \cdot \underline{n}. \tag{2.102}$$

2.11   For some continuous systems the particle velocities are so small that the kinetic energy can be neglected. This is the case, for example, when the flow of fluids through porous media is considered. In such a case the internal and total energies are equal. Furthermore, show that in cases like that, Eq. (2.66) is equivalent to

$$\rho \frac{DU}{Dt} = \nabla \cdot \left(\underline{q} + \underline{\underline{\sigma}} \cdot \underline{v} \right) + \rho \left(h + \underline{b} \cdot \underline{v} \right). \tag{2.103}$$

REFERENCES

1. Allen, M.B., I. Herrera, and G.F. Pinder, *Numerical Modeling in Science and Engineering,* Wiley, New York, 1988.

2. Atanackovic, T.M. and G. Ardeshir, *Theory of Elasticity for Scientists and Engineers*, Springer-Verlag, Berlin, 2000.

3. Atkin, R.J., *An Introduction to the Theory of Elasticity*, Longman, London, 1980.

4. Ciarlet, P.G., *The Finite Element Method for Elliptic Problems*, SIAM CL, 40, 1978.

5. Ciarlet, P.G., *Mathematical Elasticity*, Vol. II: Theory of Plates.North-Holland, Elsevier, Amsterdam, 1997.

6. Ciarlet, P.G., *Élasticité Tridimensionnelle*, Collection RMA1, 1986.

7. Ciarlet, P.G., *Mathematical Elasticity*, Vol. II: Theory of Plates, North-Holland, Elsevier, 1997.

8. Coleman, B., M.E. Gurtin, I. Herrera, and C. Truesdell, *Wave Propagation in Dissipative Materials*, Springer-Verlag, Berlin, 1965.

9. Courant, R. and D. Hilbert, *Methods of Mathematical Physics*, Vol. I, Interscience, New York, 1953.

10. Emanuel, G., *Analytical Fluid Dynamics*, CRC Press, Boca Raton, FL, 2001.

11. Eringen, A.C. and E.S. Suhubi, *Elastodynamics*, Elsevier Science,Amsterdam, 1974.

12. Eringen, A.C., *Mechanics of Continua*, 2nd ed., Krieger, Huntington, NY, 1980.

13. Eringen, A.C. and R.E. Craine, Continuum theories of mixtures: basic theory and historical development, Quart. *J. Mech. Appl. Math.*, 29(2), 209-244, 1976.

14. Ewing, W.M., W.S. Jardewsky and F. Press, *Elastic Waves in Layered Media*, McGraw-Hill, New York, 1957.

15. Fetter, C.W., *Applied Hydrogeology*, Prentice Hall, Upper Saddle River, NJ, 1994, 2001.

16. Granger, R.A., *Fluid Mechanics*, Dover Classics of Science and Mathematics, Dover, New York, 1995.

17. Green, A. E. and W. Green-Zerna, *Theoretical Elasticity*, reprint edition, Dover, New York, 1992.

18. Gurtin, M.E., The Linear Theory of Elasticity, *Handbuch der Physik*, Ed. S. Flügge, Vol. VIa/2, Springer-Verlag, Berlin, 1972.

19. Gurtin, M.E., *An Introduction to Continuum Mechanics*, Academic Press, New York, 1981.

20. Gurtin, M.E., *Topics in Finite Elasticity*, SIAM, Philadelphia, 1981.

21. Hill, R., Discontinuity relations in mechanics of solids, *Progress in Solid Mechanics*, Vol. II(72), 245-276, North-Holland, Amsterdam 1961.

22. Hunter, S.C., *Mechanics of Continuous Media*, Ellis Horwood, Chichester, UK, 1976.

23. Hutter, K. and K. Jöhnk, *Continuum Methods of Physical Modeling*, Springer-Verlag, Berlin, 2004.

24. Jeffrey, A., A note on the derivation of the discontinuity conditions across contact discontinuities, shocks and phase fronts, *Z. Angew. Math. Phys.*, 15, 68-71, 1964.

25. Keller, H.B., Propagation of stress discontinuities in inhomogeneous elastic media, *SIAM Rev.*, 6, 356-382, 1964.

26. Landau, L.D. and F.M. Lifschitz, *Theory of Elasticity*, Pergamon Press, London, 1959.

27. Lions, J.L. and E. Magenes, *Non-homogeneous Boundary Value Problems and Applications*, Springer-Verlag, Berlin (3 volumes), 1972.

28. Malvern, L.E., *Introduction to the Mechanics of a Continuous Medium,* facsimile edition, Prentice-Hall, 1977.

29. Marsden, J. E. and T.R.J. Hughes, *Mathematical Foundations of Elasticity*, reprint edition, Dover, New York, 1994.

30. McDougall, F.H., *Thermodynamics and Chemistry*, Wiley, New York, 1939.

31. Meyer, R.E., *Introduction to Mathematical Fluid Dynamics*, Dover, New York, 1982 .

32. Noll, W., A mathematical theory of the mechanical behavoir of continuous media, Archi. Rational Mech. Anal., 2, 197-226, 1958.

33. Noll, W., *The Foundations of Mechanics and Thermodynamics* (selected papers by W. Noll, with a preface by C. Truesdell), Springer-Verlag, Berlin, 1974.

34. Oliver, X. and C. Agelet, *Mecánica de Medios Continuos para Ingenieros*, Alfaomega, Edicions UPC, S.L. Universitat Politècnica de Catalunya, 2005.

35. Quarteroni, A. and A. Valli, Domain decomposition methods for partial differential equations, *Numerical Mathematics and Scientific Computation*, Oxford Science Publications, Oxford, UK, 1999.

36. Quarteroni, A., R. Sacco, and F. Saleri, *Numerical Mathematics*, Texts in Applied Mathematics, Vol. 37, Springer-Verlag, New York, 2000.

37. Smith, J.M., H.C. Van Ness, and M. Abbott, *Introduction to Chemical Engineering Thermodynamics*, McGraw-Hill, New York, 2000.

38. Sokolnikoff, I.S., *Mathematical Theory of Elasticity,* McGraw-Hill, New York, 1956.

39. Spencer, A.J.M., *Continuum Mechanics,* Longman, London, 1980.

40. Street, R.L., *The Analysis and Solution of Partial Differential Equations*, Brooks/Cole, Monterey, CA, 1973.

41. Taylor, M.E., *Partial Differential Equations: Basic Theory*, Texts in Applied Mathematics, Vol. 23, Springer-Verlag, New York, 1996.

42. Temam, R., *Navier-Stokes Equations: Theory and Numerical Analysis*, AMS Chelsea Publishing, American Mathematical Society, Providence, RI, 1984.

43. Thomas, J.W., *Numerical Partial Differential Equations: Finite Difference Methods*, Texts in Applied Mathematics, Vol. 22, Springer-Verlag, New York, 1995.

44. Thomas, J.W., *Numerical Partial Differential Equations: Conservation Laws and Elliptic Equations*, Texts in Applied Mathematics, Vol 33, Springer-Verlag, New York, 1999.

45. Toselli, A. and O. Widlund, *Domain Decomposition Methods: Algorithms and Theory*, Springer Series in Computational Mathematics, Springer-Verlag, Berlin, 2005.

46. Treves, F., *Basic Linear Partial Differential Equations*, Academic Press, New York, 1975.

47. Truesdell, C. and R.A. Toupin, The classical field theories, *Handbuch der Physik,* Vol. 3, Ed. S. FlüggeSpringer-Verlag, Berlin, 1960.

48. Truesdell, C., The rational mechanics of materials: -past, present, future, *Appl. Mech. Rev.* 12, 75-80, 1959. Reprinted in *Applied Mechanics Surveys,* Vol. 22, Spartan Books, Washington, DC, 1965.

49. Truesdell, C., General and exact theory of waves in finite elastic strain, *Arch. Rational Mech. Anal.*, 8, 263-296, 1961. Reprinted in B. Coleman, M.E. Gurtin, I. Herrera, and C. Truesdell, *Wave Propagation in Dissipative Materials*, Springer-Verlag, Berlin, 1965.

50. Truesdell, C. and W. Noll, The non-linear field theories of mechanics, in Vols. 12, 16, 19, 20, 21, 23, and 73 of the *Handbuch der Physik*, Ed. S. Flügge. Springer-Verlag, Berlin, 1965.

51. Truesdell, C., *The Elements of Continuum Mechanics*, Vol. 27, Springer-Verlag, Berlin, 1966.

52. Truesdell, C., A program toward rediscovering the rational mechanics of the age of reason, *Arch. Hist. Exact Sci.* 1, 3-36, 1960. Corrected reprint in *Essays in the History of Mechanics,* Vol. 22, Springer-Verlag, Berlin, 1968.

53. Truesdell, C., *A First Course in Rational Continuum Mechanics,* Vol. 1, Academic Press, New York,1977.

54. Wang, C.C., *Mathematical Principles of Mechanics and Electromagnetism*, Plenum Press, New York, 1979.

55. Warsi, Z.U.A., *Fluid Dynamics: Theoretical and Computational Approaches*, 2nd ed. CRC Press, Boca Raton, FL, 1998.

CHAPTER 3

MECHANICS OF NON-CLASSICAL CONTINUOUS SYSTEMS

In this chapter we will introduce several physical systems of practical importance which we will consider in more detail in later chapters. The choices are made based upon a balance between complexity and practical relevance.

3.1 MULTIPHASE SYSTEMS

A brief discussion of the concepts of mixtures and solutions, which are generally well understood, may help to better understand the differences between one-phase and multiphase systems. For example, if salt (sodium chloride) is dissolved in water, at the beginning only one liquid phase is observed. However, if we continue to increase the concentration of salt, at one point we start to observe also a solid phase, attributable to the fact that the concentration of salt has exceeded its *solubility limit* for the temperature and pressure of its environment. Then one can try to make the solid phase disappear, using, for example, a mixing device such as a rotating magnet in a beaker, until a relatively homogeneous mixture of solid and liquid (suspension) is formed. However, if we leave the mixture at rest for some time, the solid phase separates again and moves to the bottom of the container.

Mathematical Modeling in Science and Engineering: An Axiomatic Approach.
By Ismael Herrera and George F. Pinder Copyright © 2012 John Wiley & Sons, Inc.

The separation of the phases from each other in this simple experiment illustrates the fact that each of the phases has moved with a different velocity. The observation that each phase moves with its own velocity is a conspicuous and distinguishing feature of multiphase systems (that is, systems with more than one phase); in general, at each point of the domain occupied by the multiphase system, several particle velocities are defined; one for each phase.

It should be evident that the theory which we have developed up to now applies to one-phase systems only, since we have assumed that at each point of the continuous system there is defined one and only one particle velocity. However, the only modification that is required to extend such a theory to multiphase systems is to assume that more than one particle velocity may be defined, one for each phase. Many important models of science and engineering correspond to one-phase systems; however, there are many other important systems that are defined by multiple phases, such as petroleum reservoirs and geothermal fields. Thus, the extension of the theory to multiphase systems, although rather simple, is quite significant.

3.2   THE BASIC MATHEMATICAL MODEL OF MULTIPHASE SYSTEMS

The general procedure for constructing multiphase models is obtained by modifying slightly the procedure explained in the preceding chapter. The modification leads to a three-step procedure:

1. A family of extensive properties is identified;

2. Each extensive property of the family is associated with one of the phases of the system; and

3. The balance conditions are applied to each extensive property of the family, using the *particle velocity of the corresponding phase*.

Application of this strategy to the basic mathematical model of the multiphase system leads to the following system of differential equations:

$$\frac{\partial \psi^\alpha}{\partial t} + \nabla \cdot (\underline{v}^\alpha \psi^\alpha) = \nabla \cdot \underline{\tau}^\alpha + g^\alpha, \quad \forall \alpha = 1, ..., N \tag{3.1}$$

and the jump conditions

$$[\![\psi^\alpha (\underline{v}^\alpha - \underline{v}_\Sigma) - \underline{\tau}^\alpha]\!] \cdot \underline{n} = 0, \quad \forall \alpha = 1, ..., N. \tag{3.2}$$

One must now define the constitutive equations and auxiliary (boundary and initial) conditions as in Section 2.2. This complementary information about the systems plays a very important role since it is the means by which scientific and technological knowledge about a specific system is incorporated into the model. It is the intellectual glue that binds the general relationships, such as those presented in Chapter 2, to a specific physical system.

3.3 SOLUTE TRANSPORT IN A FREE FLUID

In this book, fluid motions under two sets of conditions will be discussed:

1. Flow of *free fluids*: that is, fluids that flow without restriction and thus occupy all of a contained physical space; and

2. Flow of fluids in a porous medium; that is fluids that are restricted to move in the voids of a porous material.

In this section, the basic mathematical models for the transport of dissolved substances (*solutes*) by a free fluid when it moves are derived and briefly discussed. Chapter 4 is devoted to a more thorough treatment of this kind of model.

It is important when considering the mechanics of continua that the concept of a particle be carefully defined. A *particle* in the context of transport is not a molecule; it is an assemblage of molecules such that the average behavior of the assemblage yields physical-chemical characteristics consistent with those we observe at the macroscopic level, that is, the level at which most bench-scale experiments are conducted. Thus there is molecular movement within a particle or even between particles, but our interest lies in how the average behavior of the particles within an *elementary volume* behaves.

To build the transport model we apply the general procedure explained in Section 3.2. A basic assumption on which the transport model is built is that the solute is thoroughly dissolved. Given this assumption, the solute, together with the fluid in which it is dissolved, constitutes a single phase. Therefore, the solute particles move with the velocity of the fluid particles. At every point of the physical space there is defined only one particle velocity; that is, the system is a one-phase system. On the other hand, the family of extensive properties of the model shrinks to only one property: the *mass of the solute*. The mass of the solute is an extensive property and it is expressed as an integral over the physical space occupied by any body of fluid; that is,

$$M_S(t) \equiv \int_{B(t)} c(\underline{x}, t) \, d\underline{x}. \tag{3.3}$$

The integrand in Eq. (3.3) is the intensive property associated with the solute mass and, according to Eq. (1.21), it is given by

$$c(\underline{x}, t) \equiv \lim_{\text{Vol} \to 0} \frac{M_S(t)}{\text{Vol}}. \tag{3.4}$$

So the intensive property associated with the mass of the solute is the *mass of solute per unit volume of the fluid* or, equivalently, of the solution, which is usually called the *concentration* of the solute.

According to Eq. (3.1), the system of partial differential equations governing the transport of a solute dissolved in a free fluid consists of only one equation:

$$\frac{\partial c}{\partial t} + \nabla \cdot (\underline{v}c) = \nabla \cdot \underline{\tau} + g \tag{3.5}$$

where g *is normally a chemical reaction term* (for example, radioactive decay) and $\underline{\tau}$ *is a mass flux term* generated by virtue of molecular motion; the mass flux is normally defined using *Fick's first law of diffusion* (a constitutive relationship). Note that the flux $\underline{\tau}$ exists because the macroscopic variable $c(\underline{x}, t)$ is an average of particles in a *representative elementary volume* (see Chapter 1). It accounts for behavior at the molecular level when considering properties such as $c(\underline{x}, t)$ at the macroscopic level. This induces, at the macroscopic level, diffusive behavior that is described by $\underline{\tau}$.

When discontinuities occur in the model they are governed by

$$[\![c(\underline{v} - \underline{v}_\Sigma) - \underline{\tau}]\!] \cdot \underline{n} = 0. \tag{3.6}$$

In fluids under ordinary conditions, discontinuities do not occur, so the latter equation does not need to be applied. However, if we were to study the transport of a solute across a shock wave, such as those occurring in supersonic flow, it would be necessary to incorporate Eq. (3.6) into our model.

As stated earlier, the velocity of solute particles equals that of the fluid particles; so in Eqs. (3.5) and (3.6), $\underline{v}$ is the *particle velocity* of the fluid. The *external supply*, g, may differ from zero for several reasons: for example, solute may be added while the transport phenomenon is occurring; chemical reactions may produce the solute; or, if the solute is radioactive, then $g < 0$, since it is decaying. In general, when $g = 0$ the transport process is said to be *conservative*. In *solute transport* studies $\underline{\tau}$ is said to be a *diffusive flux* and for transport in free-fluids its most significant contribution is *molecular diffusion*, which is due to *Brownian motion* of the microscopic solute particles. This phenomenon is further explained in Section 4.2.

As for the *constitutive equations* and the complementary information about the transport system that is needed to complete the model, the following points should be mentioned. For Eq. (3.5) to be a *complete model* it is necessary that the velocity of the fluid be known. For example, if the problem is one of atmospheric pollution, the velocity information could be supplied by a net of wind-measuring stations, and similarly, if transport of pollutants in a water stream is being modeled, water velocity measurements could be obtained. The information about g may be given by constitutive equations derived from physical or chemical laws, such as the laws of nuclear decay mentioned earlier, chemical balances, or chemical kinetics. In quiescent or very slowly moving fluids (gases or liquids) the constitutive equation relating the diffusive flux to the distribution of the concentration used most frequently is Fick's law, as discussed in Chapter 4. For fluids exhibiting non-uniform flow, the flux relationship may be more complex and will also be considered in a later chapter.

Before finishing this section, we note that for steady state, the transport equation, Eq. (3.5), reduces to

$$\nabla \cdot (\underline{v}c) = \nabla \cdot \underline{\tau} + g. \tag{3.7}$$

To be able to predict the distribution of the concentration and its evolution, using this equation or Eq. (3.5), the equation needs to be complemented with suitable initial and boundary conditions, which will be discussed in a later section.

3.4   TRANSPORT BY FLUIDS IN POROUS MEDIA

A *porous medium,* as used in this book, is a solid material that contains voids in its interior that are interconnected such that the flow of fluid through them is possible (see Fig. 1.3 on page 15). The solid material that forms the porous medium is referred to as the *solid matrix,* the voids are the *pores,* and the volume fraction of the physical space that is occupied by the pores is the *porosity.* In single-fluid phase systems, it is assumed that the porous medium is *saturated* by the fluid; by this we mean that the fluid fills the pores completely, and therefore the volume of the pores equals the volume of the fluid that is contained in the porous medium (see Fig. 3.1).

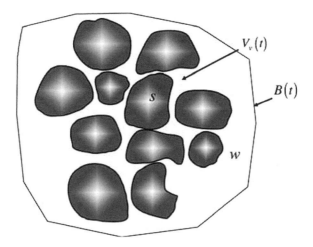

Figure 3.1   Repeat of the diagrammatic representation of a porous medium presented earlier. The medium is made up of solid s and water w.

However, the volume of a body of fluid is *not* equal to the volume of the domain of the physical space where it resides; instead, it equals the volume of the pores contained in such a domain. Clearly, the volume of the fluid is an extensive property and can be expressed as an integral: namely,

$$V_f(t) = V_p(t) \equiv \int_{B(t)} \varepsilon(\underline{x}, t) \, d\underline{x}. \tag{3.8}$$

The integrand of the integral occurring in this equation is the intensive property associated with the fluid-volume. According to Eq. (1.21), on page 8, it is given by

$$\varepsilon(\underline{x}, t) \equiv \frac{V_p}{\text{Total Vol}}. \tag{3.9}$$

Then we conclude that the intensive property, $\varepsilon(\underline{x}, t)$, associated with the fluid volume is the porosity. In Eq. (3.9), the porosity $\varepsilon(\underline{x}, t)$ is a function of both position $\underline{x}$ and time t; if the porous material is heterogeneous, ε varies with position, and generally, when the material is elastic (deformable), it varies with time.

In turn, if the fluid contains a solute, the mass of solute per unit fluid volume is again referred to as the concentration. Generally, the principal goal of a transport model is to predict the evolution of the spatial distribution of the concentration as the fluid moves. The fluid motion takes place in the pores, and the fluid particle velocity is $\underline{v}(\underline{x}, t)$. The transport models to be described in this section are based on only one extensive property: the mass of solute. The mass of solute contained in a body of fluid is also an extensive property; that is,

$$M_S(t) \equiv \int_{B(t)} \psi(\underline{x}, t) \, d\underline{x}. \tag{3.10}$$

Here $\psi(\underline{x}, t)$ is the intensive property associated with the solute mass, whose explicit expression is established next. Again applying Eq. (1.21), it is seen that

$$\psi(\underline{x}, t) \equiv \frac{M_S}{\text{Total Vol}} = \frac{V_f}{\text{Total Vol}} \frac{M_S}{V_f} = \frac{V_f}{\text{Total Vol}} \frac{M_S}{V_f} = \varepsilon(\underline{x}, t) c(\underline{x}, t). \tag{3.11}$$

Thus, in this case the intensive property associated with the solute mass is equal to the porosity multiplied by the concentration. Substituting this result in Eq. (3.10), we get

$$M_S(t) \equiv \int_{B(t)} \varepsilon(\underline{x}, t) c(\underline{x}, t) \, dx. \tag{3.12}$$

The basic mathematical model for the transport of a solute by a fluid contained in a porous medium follows immediately from Eq. (3.1), since the only intensive property of the model is εc. The differential equation is

$$\frac{\partial \varepsilon c}{\partial t} + \nabla \cdot (\varepsilon c \underline{v}) = \nabla \cdot \underline{\tau} + g \tag{3.13}$$

while the jump condition is

$$[\![\varepsilon c (\underline{v} - \underline{v}_\Sigma) - \underline{\tau}]\!] \cdot \underline{n} = 0. \tag{3.14}$$

In order for Eqs. (3.13) and (3.14) to define a complete model, the fluid motion in the porous medium needs to be known in advance; we notice that in the case of transport in a porous medium, knowing the motion of the fluid may require knowing not only the particle velocity of the fluid but also the porosity as a function of position and time. In practice, one frequently derives these functions by modeling the flow of the fluid before modeling the solute transport. The modeling of fluid flow of through porous media is presented in Chapter 5. Another possible course of action is to carry out field measurements of these variables, but except for simple cases, such measurements are generally more difficult and complicated than in the case of transport by free fluids, when only knowledge of the fluid velocity is required. The complementary information needed for the external supply g is essentially the same as that needed in solute transport by free fluids, except that when the fluid is contained in a porous medium there may be exchange of solute with the solid matrix, which corresponds to chemical reactions and/or *adsorption*. For the kind of solute transport treated in this section, the random processes that contribute to the diffusive flux stem from two sources: molecular Brownian motion and the random character of the pores and channels that constitute the solid matrix. So the diffusive flux $\underline{\tau}$ is the sum of two terms, the molecular diffusion and the *mechanical dispersion*, for which Fick's law adopts a special form; this will be discussed in Section 6.3 on page 121.

3.5 FLOW OF FLUIDS THROUGH POROUS MEDIA

The following assumptions are usually adopted for models of fluid flow through porous media:

1. The porous medium is saturated by the fluid; and

2. The mass of the fluid is conserved.

Generally, the final goal of this kind of model is to predict the flow of the fluid (that is, the fluid velocity). The model is based on only one extensive property: the fluid mass. This is essentially given by Eq. (3.12), except that the solute concentration has to be replaced by the mass of fluid per unit volume of fluid (that is, the *fluid density*). Such a replacement in Eq. (3.12) yields

$$M_f\left(t\right) \equiv \int_{B(t)} \varepsilon\left(\underline{x}, t\right) \rho\left(\underline{x}, t\right) dx \tag{3.15}$$

where $\rho\left(\underline{x}, t\right)$ is the fluid density. Clearly, the intensive property associated with the sole extensive property of this model is the product $\varepsilon\left(\underline{x}, t\right) \rho\left(\underline{x}, t\right)$. Therefore, the basic mathematical model for *flow of a fluid through a porous medium* is

$$\frac{\partial \varepsilon \rho}{\partial t} + \nabla \cdot \left(\varepsilon \rho \underline{v}\right) = g + \nabla \cdot \underline{\tau}. \tag{3.16}$$

However, since the fluid is not subjected to diffusion because there are no separate species ($\underline{\tau} = 0$) and assuming that no mass of fluid is generated in the system constituted by the porous material and the fluid (that is, $g = 0$), Eq. (3.16) reduces to

$$\frac{\partial \varepsilon \rho}{\partial t} + \nabla \cdot \left(\varepsilon \rho \underline{v}\right) = 0. \tag{3.17}$$

As for the jump condition of Eq. (3.14), it reduces to

$$[\![\varepsilon \rho \left(\underline{v} - \underline{v}_\Sigma\right)]\!] \cdot \underline{n} = 0. \tag{3.18}$$

It must be pointed out that in regional groundwater studies extraction, or injection of water by wells is frequently incorporated as a distributed *external supply*, in which case $g \neq 0$ and Eq. (3.17) is replaced by

$$\frac{\partial \varepsilon \rho}{\partial t} + \nabla \cdot (\varepsilon \rho \underline{v}) = g. \tag{3.19}$$

An important difference between fluid flow and solute-transport models is that the particle velocity is not given in the fluid-flow equation as a datum, but instead, is derived using a constitutive equation, typically *Darcy's law*, which relates the fluid velocity to the fluid-pressure spatial distribution. The substitution of Darcy's law into Eq. (3.19) yields an equation with the velocity represented via the state variable pressure (or *hydraulic head*). When other constitutive relationships are used to describe changes in ε and ρ in terms of the same state variable as that used for velocity, the final equation has only one unknown state variable.

3.6   PETROLEUM RESERVOIRS: THE BLACK-OIL MODEL

Petroleum reservoirs are conceptualized as a porous medium, usually a rock of sedimentary origin, whose pores store fluids containing hydrocarbons; when the pores are full of such fluids, the reservoir is said to be saturated. In what follows we treat only saturated reservoirs, which are the most important in petroleum engineering. Generally, the pores of a petroleum reservoir are occupied by three phases: water, liquid oil, and gas, referred to as the *water phase*, the *oil phase,* and the *gas phase,* respectively. Of these, the first two are liquids while the third is gaseous. In the pores, they are separated from each other by *interfaces*. Therefore, each phase occupies its own volume and the symbols S_w, S_o, and S_g, respectively, will be used to represent the fraction of the pore volume occupied by each of the phases, that is,

$$S_\alpha = \frac{V_\alpha}{V_p}, \quad \alpha = o, w, g$$

and V_p is the volume of void space: namely,

$$V_p = V_o + V_w + V_g.$$

Clearly, when the reservoir contains only the three fluid phases, water, oil, and gas, one has

$$S_w + S_o + S_g = 1. \tag{3.20}$$

The water that constitutes the water phase may contain different solutes, but generally the concentration of hydrocarbons in solution are ignored since petroleum is not very soluble in water. The hydrocarbons contained in the oil phase will be classified into two groups: *dissolved gas (gas in the oil phase)* and *non-volatile oil (oil in the oil phase)*. An important property implicit in this classification is whether or not the hydrocarbon in the oil phase is gaseous at atmospheric pressure. On the other hand, the hydrocarbons contained in the gas phase are necessarily gaseous at atmospheric pressure, since it will always be assumed that the reservoir pressure is above atmospheric pressure.

3.6.1 Assumptions of the black-oil model

The main objective of the black-oil petroleum-reservoir model is to predict the behavior of the fluids contained in the reservoirs; this includes the distribution in space and time of the following variables: pressure and saturation in each phase and the fractions of the *oil-phase* mass that correspond to dissolved gas and to non-volatile oil. The solid-matrix behavior is not included among the main concerns of the black-oil models. Albeit such models give some rudimentary predictions of the land subsidence that occurs when oil fields are exploited, such predictions are only an indirect by-product. More complete models for solid-matrix behavior are based on poroelasticity, a topic that is not covered in this book. Readers interested in poroelasticity are referred to reference Biot, 1941[2].

The basic hypotheses of the black-oil model are (see, for example,[references [1] and Chen et al.[3]):

1. There are three fluid phases in the reservoir: the water phase, the oil phase and the gas phase, (see Fig. 3.2);

2. Solubility of hydrocarbons in the water phase is negligible and can be ignored in predicting fluid behavior;

3. The oil phase contains dissolved gas (gas in the oil phase) and non-volatile oil (oil in the oil phase);

4. There is exchange of mass between the oil phase and the gas phase; namely, dissolved gas may go as gas into the gas phase, and conversely, gas from the gas phase may dissolve in the oil phase; and

5. In the fluids of the reservoir, mass-diffusive processes are neglected.

3.6.2 Notation

The hypotheses just listed have implications that will be used in this subsection. The water phase consists of only one component, water, and the gas phase also consists of only one component, volatile hydrocarbon. The oil phase, on the other hand, is made of two components: dissolved gas (gas in the oil phase) and non-volatile oil (oil in oil phase). In our conceptualization we have four extensive properties, but only three components. What happens is that we associate an extensive property with each component-phase pair (where such a component occurs).

The *oil-phase density*, ρ_o, is defined to be equal to the mass of the hydrocarbons contained in the oil phase per unit volume of that phase. It can be expressed as [1], [3]

$$\rho_o = \rho_{Oo} + \rho_{Go}. \tag{3.21}$$

Here ρ_{Oo} and ρ_{Go} are the *net densities* of the two components that form the oil phase: the oil in the oil phase and the gas in the oil phase, respectively. They are defined by

$$\rho_{Oo} = \frac{m_{Oo}}{V_o} \quad \text{and} \quad \rho_{Go} = \frac{m_{Go}}{V_o}. \tag{3.22}$$

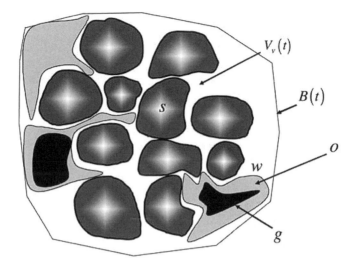

Figure 3.2   Porous medium with gas (g), oil (o), water (w), and solid (s) phases.

In this equation, m_{Oo} and m_{Go} are the masses of non-volatile oil and dissolved gas, respectively, contained in a volume V_o of the oil phase. The solution *gas:oil ratio* is defined to be

$$R_S \equiv \frac{\rho_{Go}}{\rho_{Oo}}. \tag{3.23}$$

We notice that this definition differs slightly from that given in reference [1], where the *gas:oil ratio* is defined as our R_S multiplied by ρ_{gSTC}/ρ_{oSTC} (where $\rho_{\alpha STC}$ is the density of phase α at storage tank conditions, STC). However, the definition given in Eq. (3.23) simplifies the expressions that will be used in the discussions that follow.

3.6.3 Family of extensive properties

The black-oil model is based on the family of four extensive properties that are the masses of each of the three components identified above. They are:

1. Mass of the water in the water phase;

2. Mass of the non-volatile oil contained in the oil phase: oil in the oil phase;

3. Mass of the dissolved gas contained in the oil phase: gas in the oil phase; and

4. Mass of the gas in the gas phase.

3.6.4   Differential equations and jump conditions

A simple computation shows that the above-defined four masses are given by

$$
\begin{aligned}
M_w\left(t\right) &\equiv \int_{B_w(t)} \varepsilon S_w \rho_w \, d\underline{x} \\
M_{Oo}\left(t\right) &\equiv \int_{B_{pl}(t)} \varepsilon S_o \rho_{Oo} \, d\underline{x} \\
M_{Go}\left(t\right) &\equiv \int_{B_{pl}(t)} \varepsilon S_o \rho_{Go} \, d\underline{x} \\
M_g\left(t\right) &\equiv \int_{B_g(t)} \varepsilon S_g \rho_g \, d\underline{x}.
\end{aligned}
\tag{3.24}
$$

To each of these extensive properties we apply the general equation of global balance [Eq. (1.25) on page 10]:

$$
\frac{dE}{dt}(t) = \int_{B(t)} g\left(\underline{x}, t\right) d\underline{x} + \int_{\partial B(t)} \underline{\tau}\left(\underline{x}, t\right) \cdot \underline{n}\left(\underline{x}, t\right) d\underline{x}.
\tag{3.25}
$$

We observe that for each of these four extensive properties, $\underline{\tau} = 0$, since the black-oil model assumes that no diffusive or dispersive processes are present. Hence, the global balances for these properties reduce to

$$
\left.
\begin{aligned}
\frac{dM_w}{dt}\left(t\right) &\equiv \int_{B_w(t)} g_w \, d\underline{x} \\[2mm]
\frac{dM_{Oo}}{dt}\left(t\right) &\equiv \int_{B_o(t)} g_{Oo} \, d\underline{x} \\[2mm]
\frac{dM_{Go}}{dt}\left(t\right) &\equiv \int_{B_o(t)} g_{Go} \, d\underline{x} \\[2mm]
\frac{dM_g}{dt}\left(t\right) &\equiv \int_{B_g(t)} g_g \, d\underline{x}.
\end{aligned}
\right\}
\tag{3.26}
$$

Here, we have used the notation g_γ for the external supplies of each of these extensive properties. In practice, the effect of pumping wells is frequently incorporated as an external supply. However, to simplify the presentation a little, we will not consider pumping at this stage. Then $g_w = 0$ and $g_{Oo} = 0$, necessarily, since those *components* do not exchange mass with any other component. As for g_{Go} and g_g, they may be different from zero only when there is exchange of mass between the oil and gas phases. Let g_g^{Go} be the rate, per unit total volume (including the fluid phases and the solid matrix), at which the gas dissolves in the *oil phase*; then $g_{Go} = g_g^{Go}$. Furthermore, we write g_{Go}^g for the rate, per unit total volume, at which the dissolved gas is liberated from the oil phase and goes into the gas phase; then $g_g = g_{Go}^g$. To assure conservation between the phases, we note that

$$
g_g^{Go} + g_{Go}^g = g_{Go} + g_g = 0.
\tag{3.27}
$$

In conclusion, the global balances of Eq. (3.26) are

$$\frac{dM_w}{dt}(t) = 0$$

$$\frac{dM_{Oo}}{dt}(t) = 0$$

$$\frac{dM_{Go}}{dt}(t) = \int_{B_o(t)} g_g^{Go} \, d\underline{x}$$

$$\frac{dM_g}{dt}(t) = \int_{B_g(t)} g_{Go}^g \, d\underline{x}.$$

$$(3.28)$$

In view of Eq. (3.24), the family of intensive properties that corresponds to the family of extensive properties under study is

$$\psi^w \equiv \varepsilon S_w \rho_w, \quad \psi^{Oo} \equiv \varepsilon S_o \rho_{Oo}, \quad \psi^{Go} \equiv \varepsilon S_o \rho_{Go}, \quad \psi^g \equiv \varepsilon S_g \rho_g. \tag{3.29}$$

Therefore, when the equations of local balance [Eqs. (3.1) and (3.2 on page 46] are applied to this family of intensive properties, the following system of equations is obtained:

$$\frac{\partial \varepsilon S_w \rho_w}{\partial t} + \nabla \cdot (\varepsilon S_w \rho_w \underline{v}^w) = 0 \tag{3.30}$$

$$\frac{\partial \varepsilon S_o \rho_{Oo}}{\partial t} + \nabla \cdot (\varepsilon S_o \rho_{Oo} \underline{v}^o) = 0 \tag{3.31}$$

$$\frac{\partial \varepsilon S_o R_S \rho_{Oo}}{\partial t} + \nabla \cdot (\varepsilon S_o R_S \rho_{Oo} \underline{v}^o) = g_g^{Go} \tag{3.32}$$

$$\frac{\partial \varepsilon S_g \rho_g}{\partial t} + \nabla \cdot (\varepsilon S_g \rho_g \underline{v}^g) = g_{Go}^g \tag{3.33}$$

together with the following jump conditions:

$$[[\varepsilon S_w \rho_w (\underline{v}^w - \underline{v}_\Sigma)]] \cdot \underline{n} = 0 \tag{3.34}$$

$$[[\varepsilon S_o \rho_{Oo} (\underline{v}^o - \underline{v}_\Sigma)]] \cdot \underline{n} = 0 \tag{3.35}$$

$$[[\varepsilon S_o R_S \rho_{Oo} (\underline{v}^o - \underline{v}_\Sigma)]] \cdot \underline{n} = 0 \tag{3.36}$$

$$[[\varepsilon S_g \rho_g (\underline{v}^g - \underline{v}_\Sigma)]] \cdot \underline{n} = 0. \tag{3.37}$$

where $\underline{v}_\Sigma$ is the velocity of the shock. Addition of Eqs. (3.32) and (3.33) yields

$$\frac{\partial \varepsilon (S_o R_S \rho_{Oo} + S_g \rho_g)}{\partial t} + \nabla \cdot \{\varepsilon (S_o R_S \rho_{Oo} \underline{v}^o + S_g \rho_g \underline{v}^g)\} = 0. \tag{3.38}$$

Equations (3.30), (3.31), and (3.38) constitute the system of differential equations that are used in petroleum engineering to define the *black-oil model*. The way we have derived it here is not standard and thereby supplies an extra equation, for example, Eq. (3.33), which can be used to evaluate the amount of gas exchange between the oil and gas phases.

Finally it is important to mention that the basic mathematical model defined via Eqs. (3.30) to (3.37) is not sufficient to define a complete model capable of predicting the behavior of a petroleum reservoir. To achieve a complete model it is necessary to supplement these equations with constitutive equations that incorporate additional scientific and technological information about the specific reservoir to be modeled. Such constitutive equations will be presented in Chapter 7, where a more thorough discussion of oil reservoirs is included.

3.7 SUMMARY

The objective of this chapter is to introduce important physical systems that will be considered in greater detail later in the book. An additional goal is to illustrate the application of the axiomatic method to multiphase systems. Since single-phase flow was considered in Chapter 2, we began our discussion in this chapter with the presentation of multiphase systems. After presentation of the multiphase mathematical model, we discussed the concept of transport, first in a single-phase fluid and then in a porous medium. Having discussed transport processes in porous media, we moved on to consider flow in porous media. Finally we considered the special case of multiphase flow in porous media known as the black-oil model, which is used extensively in petroleum engineering.

EXERCISES

Equations (3.1) and (3.2) constitute a master form which can be used for constructing the basic mathematical models of many engineering and scientific systems. At first glance these equations may appear as too abstract, but they will become friendlier after some of the particular cases they contain are derived and the exercises that follow have such a purpose.

3.1   Write the basic mathematical models of each of the examples indicated.

   a) A one-phase system such that the cardinality (recall that the cardinality of a set equals the number of its members) of its characteristic family of extensive properties is 1.

   b) A three-phase system such that the cardinality of its characteristic family of extensive properties is 3.

   c) A five-phase system such that the cardinality of its characteristic family of extensive properties is 5.

3.2   For each of the systems considered in Exercise 3.1, the cardinality of its characteristic family of extensive properties equals the number of phases of the system. More complicated and interesting situations occur when this is not the case. This is illustrated in the ensuing examples. Write the basic mathematical models for each of the following specific cases:

a) A one-phase system such that the cardinality of its characteristic family of extensive properties is 2.

b) A two-phase system such that the cardinality of its characteristic family of extensive properties is 3. Assume that the first two extensive properties are associated with phase 1, while the third extensive property is associated with phase 2.

c) A three-phase system such that the cardinality of its characteristic family of extensive properties is 4. Assume that:

    i. Extensive property 1 is associated with phase 1;

    ii. Extensive properties 2 and 3 are associated with phase 2; and

    iii. Extensive property 4 is associated with phase 3.

3.3 Let N be the cardinality of the characteristic family of extensive properties of an M-*phase* continuous system (observe that $N \geq M$). Then, in general, the basic mathematical model of Eqs. (3.1) and (3.2) can be written in the form

$$\frac{\partial \psi^\alpha}{\partial t} + \nabla \cdot \left(\psi^\alpha \underline{v}^{\beta(\alpha)} \right) = \nabla \cdot \underline{\underline{\tau}}^\alpha + g^\alpha, \quad \forall \alpha = 1, ..., N \tag{3.39}$$

and

$$\left[\!\left[\psi^\alpha \left(\underline{v}^{\beta(\alpha)} - \underline{v}_\Sigma \right) - \underline{\underline{\tau}}^\alpha \right]\!\right] \cdot \underline{n} = 0, \quad \forall \alpha = 1, ..., N. \tag{3.40}$$

Here $\beta(\alpha)$ is a function defined in the set of numbers $\{1, ..., N\}$ and with values in the set $\{1, ..., M\}$.

Firstly, consider the cases when the function $\beta(\alpha)$ is given, respectively, by:

1a:

α	β
1	1

1b:

α	β
1	1
2	2
3	3

1c:

α	β
1	1
2	2
3	3
4	4
5	5

Show that the mathematical models correspond to the three cases considered Exercise 3.1.

Second, consider the cases when the function $\beta(\alpha)$ is given, respectively, by:

2a:

α	β
1	1
2	1

2b:

α	β
1	1
2	1
3	2

2c:

α	β
1	1
2	2
3	2
4	3

Show that the mathematical models described by these three tables correspond to models 3.2.a, 3.2.b and 3.2.c of Exercise 3.2, respectively.

3.4   Consider a three-phase system whose characteristic system of extensive properties has four members, for which table 2c) holds. In such a case, the system of differential equations (3.39) is

$$\frac{\partial \psi^1}{\partial t} + \nabla \cdot \left(\psi^1 \underline{v}^1 \right) = \nabla \cdot \underline{\underline{\tau}}^1 + g^1,$$

$$\frac{\partial \psi^2}{\partial t} + \nabla \cdot \left(\psi^2 \underline{v}^2 \right) = \nabla \cdot \underline{\underline{\tau}}^2 + g^2,$$

$$\frac{\partial \psi^3}{\partial t} + \nabla \cdot \left(\psi^3 \underline{v}^2 \right) = \nabla \cdot \underline{\underline{\tau}}^3 + g^3, \tag{3.41}$$

$$\frac{\partial \psi^4}{\partial t} + \nabla \cdot \left(\psi^4 \underline{v}^3 \right) = \nabla \cdot \underline{\underline{\tau}}^4 + g^4.$$

From this system of equations, derive the differential equations of the basic mathematical model of a black-oil system, Eqs.(3.30) to (3.33), taking into account that the three phases of such a system are water, liquid-oil and gas, respectively, while the intensive properties are given by Eq.(3.29); i.e.,

$$\psi^1 = \varepsilon S_w \rho_w \tag{3.42}$$
$$\psi^2 = \varepsilon S_o \rho_{Oo} \tag{3.43}$$
$$\psi^3 = \varepsilon S_o \rho_{Go} \tag{3.44}$$
$$\psi^4 = \varepsilon S_g \rho_g. \tag{3.45}$$

Explicitly state the assumptions that are required to obtain Eqs. (3.30) to (3.33).

3.5   Show that for non-diffusive and conservative $(g = 0)$ transport of a solute in an incompressible free fluid, fluid particles conserve their concentration.

Hint: First show that the condition *fluid particles conserve their concentration* is equivalent to

$$\frac{Dc}{Dt} = 0. \tag{3.46}$$

3.6   The *mass ratio* is defined to be

$$\omega \equiv \frac{c}{\rho}. \tag{3.47}$$

Assume that both the solute and the free fluid in which the solute is contained are non-diffusive and conservative. Then show that fluid particles conserve their mass ratio even if the fluid is compressible.

Hint: Combine the mass conservation equation of the solute with that for the fluid.

3.7   Show that for non-diffusive and conservative transport of a solute in an incompressible fluid contained in a porous medium, fluid particles conserve their concentration.

3.8   Show that in transport in a porous medium, the fluid particles also conserve their mass ratio when both the solute and the fluid are conservative and non-diffusive, even if the fluid is compressible.

3.9   In this exercise the student is required to prove the *bubble-point conservation principle*. The meaning of this statement is explained next.

When oil is in a tank in which a gas phase is not present, the value of the *gas:oil ratio*, R_S, will not change as we decrease the oil pressure. However, if we continue decreasing the oil pressure a point will be reached at which the oil starts to bubble and a gas phase starts to form. Such a pressure is called the *bubble-point pressure*. The bubble-point pressure depends on the value of the gas:oil ratio, and there is one-to-one correspondence between bubble-point pressure and R_S (see Fig. 3.3). The *bubble-point conservation principle* states that *in the absence of a gas phase, oil particles conserve their bubble point.*

Hint: First observe that the statements \oil particles conserve their bubble point". and \oil particles conserve R_S" are equivalent. Then show that

$$\frac{DR_S}{Dt} = 0. \tag{3.48}$$

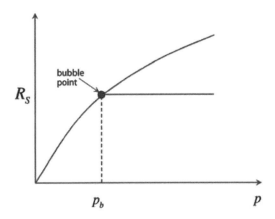

Figure 3.3   Schematic of pressure versus R_S, illustrating the concept of the *bubble point.*

3.10   Let Ω be a domain of the physical space and $\partial\Omega$ its boundary. Show that when a free fluid is incompressible, the following equation is fulfilled:

$$\int_{\partial\Omega} \underline{v} \cdot \underline{n}\, dx = \int_{\Omega} q\, dx. \tag{3.49}$$

Here q is the volume rate per unit volume at which the fluid is injected $\underline{n}$ is the unit normal vector.

3.11   When considering fluid porous systems the solid porous phase is referred to as the *solid matrix*. When the solid matrix is compressible, its porosity depends on the fluid pressure; in such a case, generally, the porosity is a function of time and we write $\varepsilon(\underline{x}, t)$. On the other hand, when the solid matrix is incompressible, the porosity is time independent and we write $\varepsilon(\underline{x})$.

Show that when the fluid and the solid matrix are incompressible,

$$\int_{\partial\Omega} \underline{U} \cdot \underline{n} \, dx = \int_{\Omega} q \, dx. \tag{3.50}$$

Here $\underline{U}$ is the *Darcy velocity*, which is defined by

$$\underline{U} \equiv \varepsilon\underline{v}. \tag{3.51}$$

Observe that the right-hand side in Eq. (3.50) equals the total volume rate at which the fluid is being injected in the physical domain Ω, while the left-hand side is the total volume rate at which the fluid is flowing through the surface of Ω. Thus, the Darcy velocity is the volume rate, per unit area, at which the fluid is flowing through the surface of Ω.

REFERENCES

1. Aziz, K. and A. Settari, *Petroleum Reservoir Simulation*, Applied Science Publishers, London, 1979.

2. Biot, M., General theory of three dimensional consolidation, J. Appl. Phys., 12, 155-164, 1941.

3. Chen, Z., G. Wang, and Y. Ma, Computational methods for multiphase flow in porous media, *Soc. Ind. and Appl. Math*, p. 549, 2006.

4. McDougall, F.H., *Thermodynamics and Chemistry*, Wiley, New York, 1939.

CHAPTER 4

SOLUTE TRANSPORT BY A FREE FLUID

An interesting phenomenon that may occur in a continuous system such as a solid or a fluid is the transport of matter and energy. This is an important subject which we will consider in this chapter. Fluids in motion carry with them energy and dissolved substances that we will denote as solutes. Furthermore, even when the continuous system is at rest, diffusion processes occur in its interior that also cause transport of the energy and the solutes that it contains. With respect to *solute transport*, in this book two kinds of models are considered: transport of solutes by free fluids and transport of solutes by fluids in porous media. This chapter is devoted to the former and Chapter 6 to the latter.

To formulate a complete model of transport, as presented in this chapter and Chapter 6, it is necessary to know in advance the *particle velocities* of the fluid as functions of position and time. A procedure that is frequently used to meet this requirement is first to calculate the particle velocities by means of flow models and, after that has been done, to utilize these particle velocities in the transport models. The flow models for free fluids are based upon the theory of fluid dynamics discussed in Chapter 10, and the flow models for fluids in a porous medium treated in Chapter 5. However, in many cases it is not possible to compute the fluid velocities by means of a flow model, in which case direct measurements of the fluid velocities need to

Mathematical Modeling in Science and Engineering: An Axiomatic Approach.
By Ismael Herrera and George F. Pinder Copyright © 2012 John Wiley & Sons, Inc.

be used. For example, when modeling contaminant transport by a river or by the atmosphere, data about the fluid velocities are obtained via measurements that are carried out at monitoring stations that are established for this purpose.

In some problems the concentration of the solutes is so small that the fluid density is essentially independent of concentration. However, in others it is so large that the density is highly dependent on the concentration, and in the latter case, when body forces such as gravity are acting on the continuous system, such density changes modify the flow conditions so that the transport and flow equations are necessarily coupled and need to be treated simultaneously. Furthermore, the resulting system is non-linear; that is, the concentration depends on the velocity and the velocity depends on the concentration in such a manner that the transport and flow equations are coupled. Something similar happens when treating energy transport; when temperature changes are large, the fluid density is strongly dependent on the temperature. Then, the transport and flow equations are once again coupled and must be solved simultaneously. In this book, only the case when the transport and flow equations are uncoupled, that is those in which the concentration and temperature are considered to have little impact on velocity, will be treated; albeit the oil-reservoir models discussed in Chapter 8, in which the transport of matter and the flow of fluids are coupled, constitute an exception to what has just been said. Whatever the case may be, the simple manner of introducing the concepts here introduced forms a necessary foundation for the mathematical modeling of more complicated systems in which coupling of the equations does occur.

4.1 THE GENERAL EQUATION OF SOLUTE TRANSPORT BY A FREE FLUID

In the study of this phenomenon, discontinuous models (shocks) will not be considered. We start by recalling some points of the discussion presented in Section 3.3 beginning on page 47. The solute mass is an extensive property given at any time t by

$$M_S(t) \equiv \int_{B(t)} c(\underline{x}, t) \, d\underline{x}. \tag{4.1}$$

In the absence of shocks the global balance equation, Eq. (2.2) on page 24, is

$$\frac{dM_S}{dt}(t) = \int_{B(t)} g_S(\underline{x}, t) \, d\underline{x} + \int_{\partial B(t)} \underline{\tau}_S(\underline{x}, t) \cdot \underline{n} \, d\underline{x}. \tag{4.2}$$

The external supply of the solute under consideration, $g_S(\underline{x}, t)$, can be different from zero for several reasons; for example, it may be attributable to radioactive decay; other examples will be discussed later in this chapter. The expression $\underline{\tau}_S(\underline{x}, t) \cdot \underline{n}$ in Eq. (4.2) represents the mass of solute per unit area, per unit time, that enters the body of fluid. In the case of a free fluid the flux of solute mass, $\underline{\tau}_S(\underline{x}, t)$, occurs mainly due to molecular diffusion, which will be discussed soon.

The differential equation of local balance is

$$\frac{\partial c}{\partial t} + \nabla \cdot (\underline{v}c) = \nabla \cdot \underline{\tau}_S + g_S \tag{4.3}$$

where $c(\underline{x}, t)$ is the solute concentration and is the intensive property associated with the solute mass, that is, the mass of solute per unit volume of solution, as defined by Eq. (3.4). As for the jump conditions of Eq. (3.6), as noted, they will not be used in the present chapter.

In order to have a complete model for the transport of a solute, Eq. (4.3) needs to be supplemented with scientific and technical information about $\underline{v}$, the particle velocity, $\underline{\tau}_S$, the solute-mass flux, and g_S, the external supply of solute mass.

4.2 TRANSPORT PROCESSES

Equation (4.3) is quite significant since it is the basis for many kinds of transport models of substances dissolved in a free fluid, such as contaminants that may occur in a surface water body or in the atmosphere. As noted earlier, Eq. (4.3) is not yet a complete model; actually, it is only what we have called the basic mathematical model, that is, the basic framework in which it is necessary to incorporate the scientific and technological information available about the processes that occur in the transport system. Three such processes will be distinguished: advection, diffusion, and mass generation.

4.2.1 Advection

Whenever the fluid is in motion, *advection* occurs; that is, whenever the particle velocity is different from zero, $\underline{v} \neq 0$. This phenomenon, or process, is due to the fact that the dissolved substance is carried by the fluid as it moves, just as a passenger is carried away by a bus. The strength of the advective process is characterized by the fluid velocity, which in the transport models that are being considered is assumed to be a datum. Thus, in order to apply Eq. (4.3) to model the transport of a solute, it is necessary to gather information about the particle velocity by suitable means.

4.2.2 Diffusion processes

The fact that a fluid is macroscopically at rest does not imply that the microscopic particles that constitute it are also at rest. As a matter of fact, the microscopic particles that constitute a fluid (atoms, ions, or molecules) are in permanent agitation, and the solute particles that accompany them perform random walks known as *Brownian motion*. Generally, the mass of solute particles that leave a fluid body due to such Brownian motion is not equal to the mass that enters it. This imbalance produces a *mass flux* across the body boundary, and in macroscopic physics, phenomena of this kind are known as *diffusion processes*. The diffusion process that is due to such a Brownian motion exclusively is known as *molecular diffusion*.

Generally, diffusion processes tend to smooth out, as time goes by, any rugosity occurring in the space distribution of the concentration. A simple experiment that evidences the occurrence of diffusion processes consists of depositing an ink drop in a pond containing water at rest. After awhile, the spot of ink spreads, getting larger and larger, and after a longer lapse, the ink concentration becomes distributed so evenly that it is imperceptible to the human eye. In free fluids, Brownian motions are always present, so that solute transport in such fluids is always diffusive. However, when diffusion processes are very small in comparison with other processes that are taking place, satisfactory results can be obtained using models that ignore them. Such transport models are said to be *non-diffusive*.

A well-known result of linear algebra shows that the most general vector-valued linear transformation of a vector can always be expressed as the product of the vector and a matrix. A very simple model, albeit widely used, for molecular diffusion is *Fick's first law*; it assumes that the vector field representing the *flux of solute mass*, $\underline{\tau}_S(\underline{x}, t)$, is a linear function of the gradient of the concentration. Hence, Fick's first law states that

$$\underline{\tau}_s(\underline{x}, t) = \underline{\underline{D}} \cdot \nabla c. \tag{4.4}$$

Here $\underline{\underline{D}}$ is a matrix, called the *tensor of molecular diffusion*, which is symmetric and non-negative. Observe also the similarity between Fick's first law and Fourier's law of Eq. (2.81) on page 37. Equation (4.4) is a very general form of Fick's first law; usually, for free fluids molecular diffusion is isotropic, in which case there is no preferred direction of the physical space with regard to the diffusion process. When the diffusion processes are isotropic, Eq. (4.4) reduces to

$$\underline{\underline{D}} \equiv D\underline{\underline{I}}. \tag{4.5}$$

Here D is a scalar known as the *diffusion coefficient*. For *isotropic diffusion,* the tensor of molecular diffusion is non-negative if and only if the diffusion coefficient is non-negative.

We observe that when $\underline{\underline{D}}$ vanishes, the assumption that the flux of solute mass is a linear function of the concentration gradient is fulfilled; thus, Fick's first law includes non-diffusive transport as a particular case. It can also be shown that the tensor of molecular diffusion is symmetric and non-negative if and only if the following condition is satisfied:

- *When the concentration gradient is non-vanishing and a surface is orthogonal to it, the flow of solute across the surface has the same sense as the gradient. At any point on the boundary of a fluid body, the inflow of solute is non-negative whenever the normal derivative of the concentration is positive.*

4.3 MASS GENERATION PROCESSES

The rate at which mass is generated is determined by the external supply, $g_S(\underline{x}, t)$. When the external supply is identically zero, no mass is generated and each body of fluid conserves the mass of solute that it contains, except for the mass that comes through its boundary due to diffusion. In this case, the transport system is said to be *conservative*.

On the other hand, the solute mass is not conserved whenever the external supply of the solute is different from zero, $g_S(\underline{x}, t) \neq 0$, and such a transport system is said to be *non-conservative*. Informally, one says that there is a *mass source* or a *mass sink* when $g_S(\underline{x}, t) > 0$ or $g_S(\underline{x}, t) < 0$, respectively. The origins of such sources and sinks are diverse; for example, two that are especially significant are radioactive decay and chemical reactions between different solutes contained in the fluid. However, the external supply of the solute may have less orthodox sources. For example, in applications in the study of contaminant transport by a river, the external supply of the contaminants may come from human activity due to communities living on the riverside.

One basic principle of physics is mass conservation. Thus, it should be observed that the fact that $g_S(\underline{x}, t) \neq 0$ does not mean that such a principle is violated, since the mass conservation principle applies to isolated systems only. The balance represented by Eq. (4.2) accounts only for the mass of the substances that are dissolved in the fluid and have precisely the chemical form of the solute. If we were going to carry out an entire mass-balance of an isolated system that contains the solute-fluid system, we would find out that the mass conservation principle is satisfied by it, because other components of the system would lose the mass that is gained by the solute-fluid system. In other words, mass that is present in the isolated system in other chemical forms may be transformed into the chemical compound that constitutes the dissolved substance. A typical example that illustrates this point is radioactive decay; when radioactive decay takes place the mass conservation principle is not violated; all that happens is that the mass of the radioactive substance that disappears is transformed into either mass of the particles that constitute radiation; namely, the *alpha* and *beta* radiation, since the *gamma* radiation is massless. .

4.4 DIFFERENTIAL EQUATIONS OF DIFFUSIVE TRANSPORT

As noted earlier, *Fick's first law* is generally accepted as the basic constitutive equation for molecular diffusion. For free fluids, diffusion processes are usually isotropic and, therefore, Eq. (4.4) will be used in what follows with the tensor of molecular diffusion given by Eq. (4.5). When this is done the differential equation of local balance, Eq. (4.3), becomes

$$\frac{\partial c}{\partial t} + \nabla \cdot (c\underline{v}) = g_S + \nabla \cdot (D\nabla c) . \tag{4.6}$$

Eq. (4.6) is a very general form of the differential equation that governs solute transport by free fluids. Some special cases of it that deserve mention are listed next:

1. If the fluid is incompressible $(\nabla \cdot \underline{v} = 0)$, one has

$$\frac{\partial c}{\partial t} + \underline{v} \cdot \nabla c = g_S + \nabla \cdot (D\nabla c) . \tag{4.7}$$

2. When the fluid is homogeneous, the diffusion coefficient is independent of position $(\nabla D = 0)$ and Eq. (4.7) becomes

$$\frac{\partial c}{\partial t} + \nabla \cdot (c\underline{v}) = g_S + D\nabla^2 c. \tag{4.8}$$

3. When the fluid is at rest $(\underline{v} = 0)$, Eq. (4.8) reduces to

$$\frac{\partial c}{\partial t} = g_S + \nabla \cdot (D\nabla c). \tag{4.9}$$

For conservative transport by a homogeneous fluid that is at rest, the governing differential equation reduces further to the well-known *heat equation*:

$$\frac{\partial c}{\partial t} = D\nabla^2 c. \tag{4.10}$$

The global balance equation of Eq. (4.2) applies to a body of fluid, that is, to a domain whose boundaries move with the fluid velocity. A balance equation that is applicable to a domain that is fixed in space can easily be derived from Eq. (4.6). Indeed, let $M_S^\Omega(t)$ be the mass of solute contained in a fixed domain, Ω, of the physical space. Then

$$M_S^\Omega(t) \equiv \int_\Omega c(\underline{x}, t)\, d\underline{x}. \tag{4.11}$$

By integration of Eq. (4.6), one gets

$$\frac{dM_S^\Omega}{dt} = \frac{d}{dt} \int_\Omega c\, d\underline{x} = \int_\Omega \{\nabla \cdot (D\nabla c - \underline{v}c) + g_S\}\, d\underline{x}. \tag{4.12}$$

Then application of the Gauss theorem yields

$$\frac{dM_S^\Omega}{dt} = \int_\Omega g_S\, d\underline{x} + \int_{\partial\Omega} \left(D\frac{\partial c}{\partial n} - c\underline{v} \cdot \underline{n}\right) d\underline{x}. \tag{4.13}$$

This equation can also be written as

$$\frac{dM_S^\Omega}{dt}(t) = \int_\Omega g_S(\underline{x}, t)\, d\underline{x} + \int_{\partial\Omega} \tau(\underline{x}, t) \cdot \underline{n}\, d\underline{x} - \int_{\partial\Omega} c\underline{v} \cdot \underline{n}\, d\underline{x}. \tag{4.14}$$

The first integral on the right-hand side of this equation is the mass of solute produced by the mass sources located in the interior of Ω. As for the second integral, it represents the mass of solute that the diffusion process carries into Ω. The term

$$\int_{\partial\Omega} c\underline{v} \cdot \underline{n}\, d\underline{x} \tag{4.15}$$

which is being subtracted, represents the mass of solute that is taken out of Ω by advection. This motivates referring to the vector field $c(\underline{x}, t)\underline{v}(\underline{x}, t)$ as the *advective flux*, just as we refer to $\tau_s(\underline{x}, t)$ as the *diffusive flux*. Then the *total mass flux* due to diffusion and advection equals $\tau_s(\underline{x}, t) - c(\underline{x}, t)\underline{v}(\underline{x}, t)$. We observe that the only difference between the global balance of Eq. (4.2) and that of Eq. (4.14) is the term due to advective flux. In conclusion, the change in solute mass contained in a domain fixed in the physical space equals the mass produced by internal sources plus the mass entering the domain due to diffusion, minus the mass taken away from the domain by advective flux.

Steady states are also of interest in many applications. The equations applicable to them follow immediately from Eqs. (4.6) to (4.10). They are:

1. The most general case, governed by

$$-\nabla \cdot (D\nabla c) + \nabla \cdot (c\underline{v}) = g_S. \tag{4.16}$$

2. The case for incompressible fluids:

$$-\nabla \cdot (D\nabla c) + \underline{v} \cdot \nabla c = g_S. \tag{4.17}$$

3. The case of a homogeneous fluid:

$$-D\nabla^2 c + \nabla \cdot (c\underline{v}) = g_S. \tag{4.18}$$

4. The situation when the fluid is at rest:

$$-\nabla \cdot (D\nabla c) = g_S. \tag{4.19}$$

5. The representation of a homogeneous fluid at rest, when the transport is conservative:

$$-\nabla^2 c = 0. \tag{4.20}$$

This last equation is the *Laplace equation*, which is the prototype for elliptic equations.

4.5 WELL-POSED PROBLEMS FOR DIFFUSIVE TRANSPORT

The mathematical studies on well-posed problems for partial differential equations refer mainly to questions of existence and uniqueness of solution. The coverage of such questions has been more thorough for linear equations than for non-linear equations. The differential equation that governs the time-dependent states of diffusive transport, Eq. (4.6) with $D > 0$, is of parabolic type. Generally, $g_S(\underline{x}, t, c)$ may be a function of position, time, and concentration. Equation (4.6) is linear unless $g_S(\underline{x}, t, c)$ is non-linear. Furthermore, it is a parabolic equation, and the well-posed problems are essentially the same as those discussed for the heat equation in Appendix A.

4.5.1   Time-dependent problems

When $g_S(\underline{x}, t, c)$ is linear, the well-posed problems for Eq. (4.6) are *initial-boundary-value problems*; that is, initial conditions are prescribed in addition to boundary conditions. Let Ω be a space domain and $\partial\Omega$ be its boundary

Initial conditions. The initial value of the concentration is prescribed:

$$c(\underline{x}, 0) = c_0(\underline{x}); \quad \forall \underline{x} \in \Omega. \tag{4.21}$$

Here the function $c_0(\underline{x})$ is the *initial concentration*, which is assumed to be a datum of the problem. Notice that although the initial time has been taken here to be $t = 0$, any other time could be chosen.

Robin boundary conditions. The most general form of boundary conditions to be considered are known as boundary conditions of *Robin type*. Let α and β be numbers such that $\alpha^2 + \beta^2 = 1$. Then

$$\alpha \frac{\partial c}{\partial n}(\underline{x}, t) + \beta c(\underline{x}, t) = \gamma(\underline{x}, t); \quad \forall \underline{x} \in \partial\Omega \text{ and } t > 0 \tag{4.22}$$

where the function $\gamma(\underline{x}, t)$ is a datum of the problem.

The Robin boundary condition, in its more general form, admits the possibility that α and β change from point to point and from time to time, that is, that α and β are functions of position and time. Thereby, we also observe that requiring $\alpha^2 + \beta^2 = 1$ is a simple manner of granting that α and β are not both simultaneously equal to zero. Two very important particular cases of *Robin boundary conditions* must be mentioned: *Dirichlet* and *Neumann boundary conditions*. They correspond to the choices $\beta = 1$ (i.e., $\alpha = 0$) and $\alpha = 1$ (i.e., $\beta = 0$), respectively.

Dirichlet boundary conditions. They are:

$$c(\underline{x}, t) = \gamma(\underline{x}, t); \quad \forall \underline{x} \in \partial\Omega \text{ and } t > 0.$$

Neumann boundary conditions. They are:

$$\frac{\partial c}{\partial n}(\underline{x}, t) = \gamma(\underline{x}, t); \quad \forall \underline{x} \in \partial\Omega \text{ and } t > 0. \tag{4.23}$$

Thus, the most general well-posed problems that will be considered when modeling diffusive transport are *initial-boundary-value problems* with Robin boundary conditions. Such problems consist of obtaining a function $c(\underline{x}, t)$ that satisfies Eq. (4.6) for every $\underline{x} \in \Omega$ and $0 \leq t \leq T$, where T is a real number greater than zero, together with the initial conditions of Eq. (4.21) and the boundary conditions of (4.22). We observe that the boundary conditions of Dirichlet type ($\beta = 1$, $\alpha = 0$) correspond to prescribing the concentration on the boundary of the region where the problem is posed, while the boundary conditions of Neumann type ($\alpha = 1$, $\beta = 0$) correspond to prescribing the diffusive flux. The latter assertion is based on Eq. (4.23) because it implies that

$$D\frac{\partial c}{\partial n}(\underline{x},t) = D\gamma(\underline{x},t); \quad \forall \underline{x} \in \partial\Omega \text{ and } t > 0. \tag{4.24}$$

Another boundary condition that is sometimes prescribed is the *total inflow of mass*, per unit area, due to diffusion and advection; that is,

$$D\frac{\partial c}{\partial n}(\underline{x},t) - (\underline{v}(\underline{x},t) \cdot \underline{n})c(\underline{x},t) = \gamma(\underline{x},t); \quad \forall \underline{x} \in \partial\Omega \text{ and } t > 0. \tag{4.25}$$

Clearly, this is a particular case of a Robin boundary condition.

In Fig. 4.1 one observes the solution to Eq. (4.6) for the one-dimensional case in which the domain of definition of the problem is the unit interval $(0, 1)$. Furthermore, $g_S = 0$ and the values of D employed are $0.001, \ 0.005,$ and 0.01. The Dirichlet boundary conditions are

$$\frac{\rho_i}{\rho} = c_i = 1, \quad x = 0 \tag{4.26}$$

$$\frac{\partial\rho_i}{\partial x} = \frac{\partial c_i}{\partial x} = 0, \quad x = 1 \tag{4.27}$$

and the initial condition is

$$\frac{\rho_i}{\rho} = c_i = 0, \quad x > 0.$$

Notice that the curves obtained are symmetrical around the concentration of 0.5, indicating that the amount of mass in the system is being conserved. The effect of increasing the value of D is to decrease the slope of the curve, thus \smearing" the solution. The location of the concentration value of 0.5 indicates the distance the solution has been convected.

4.5.2 Steady state

Well-posed time-independent problems do not include initial conditions. As for the boundary conditions, they are the same as those that have already been considered for time-dependent problems. However, when Neumann boundary conditions are imposed on the whole boundary of the domain, the solution may not exist. When the prescribed boundary data are such that the solution does exist, the data are said to be *compatible,* and in such a case the solution is non-unique; any two solutions differ by a constant.

4.6 FIRST-ORDER IRREVERSIBLE PROCESSES

For *conservative* processes, $g_S = 0$ everywhere, and Eq. (4.6) reduces to

$$\frac{\partial c}{\partial t} + \nabla \cdot (c\underline{v}) = \nabla \cdot (D\nabla c). \tag{4.28}$$

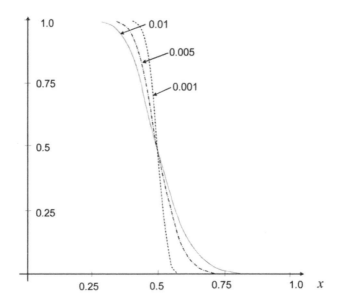

Figure 4.1   Concentration curves calculated for $v = 1$, $t = 0.5$, $g_s = 0$, and $D = 0.001$, 0.005, and 0.01.

Certain kinds of chemical activity, such as radioactive decay, hydrolysis, and certain forms of biodegradation, can be characterized as first-order irreversible processes. For them [4],[2], the following expression for the external supply applies:

$$g_S(\underline{x}, t) = -\lambda c(\underline{x}, t) \tag{4.29}$$

where λ is a positive constant. If the fluid is homogeneous, Eq. (4.6) yields

$$\frac{\partial c}{\partial t} + \nabla \cdot (c\underline{v}) + \lambda c - \nabla \cdot (D\nabla c) = 0. \tag{4.30}$$

In Fig. 4.2 are presented solutions to the problem described by Eq. (4.30) with different values of λ. The value of $D = 0.5$, $v = 1$, and the elapsed time is $t = 30$. The most evident impact on the solution of this first-order chemical reaction is the loss in overall mass which is reflected in the smaller area under each curve as the reaction rate is increased. The boundary conditions are

$$\frac{\rho_i}{\rho} = c_i = 1, \quad x = 0$$

$$\frac{\partial \rho_i}{\partial x} = \frac{\partial c_i}{\partial x} = 0, \quad x = 45$$

and the initial condition is

$$\frac{\rho_i}{\rho} = c_i = 0, \quad x > 0.$$

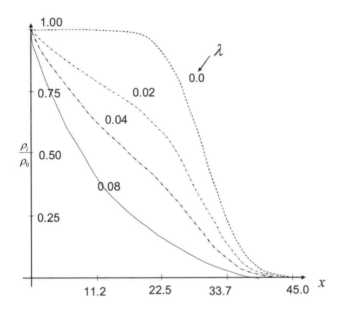

Figure 4.2   Concentration profiles using $D = 0.5$, $t = 30$, $v = 1$, and $\lambda = 0.0, 0.2, 0.4$, and 0.8.

4.7 DIFFERENTIAL EQUATIONS OF NON-DIFFUSIVE TRANSPORT

The governing equation for non-diffusive solute-transport can be derived by setting $D = 0$ in Eq. (4.6). This yields

$$\frac{\partial c}{\partial t} + \nabla \cdot (c\underline{v}) = g_S\left(c, \underline{x}, t\right). \tag{4.31}$$

The equation for the case when the fluid is incompressible is also worth noting; under these circumstances Eq. (4.31) becomes

$$\frac{\partial c}{\partial t} + \underline{v} \cdot \nabla c = g_S\left(c, \underline{x}, t\right). \tag{4.32}$$

For steady states Eq. (4.31) reduces further to

$$\nabla \cdot (c\underline{v}) = g_S\left(c, \underline{x}, t\right). \tag{4.33}$$

When the fluid is incompressible and transport is at steady state, we obtain

$$\underline{v} \cdot \nabla c = g_S\left(c, \underline{x}, t\right). \tag{4.34}$$

4.8 WELL-POSED PROBLEMS FOR NON-DIFFUSIVE TRANSPORT

A very important difference between the general governing equations of diffusive and non-diffusive transport is that Eq. (4.6) contains second-order spatial derivatives, since $D > 0$, whereas Eq. (4.31) does not. As a consequence, the equation of diffusive transport is a second-order parabolic equation, while that of non-diffusive transport is a *first-order equation*. First-order partial differential equations can be reduced to a family of first-order ordinary differential equations, each of them satisfied along certain curves called *characteristic curves* which were mentioned above (see, [[2]]); as will be seen, characteristic curves for non-diffusive transport are the trajectories, in space-time, of the fluid particles (see Fig. 4.3). Therefore, the solution of such a partial differential equation is completely determined when its value is prescribed at one, and only one, point of each space-time characteristic curve. Consequently, the well-posed problems are those that fulfill this condition and the following general statement holds.

Well-posed boundary-value problems of non-diffusive transport of a solute are those in which the value of the solute-concentration is prescribed at one, and only one, point of the space-time trajectory of each fluid particle.

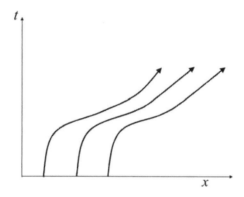

Figure 4.3   Trajectories of fluid particles are characteristic curves.

4.8.1   Well-posed problems in one spatial dimension

The concepts to be presented below are easier to understand when only one spatial dimension is involved; thus, we begin our analysis with that case. The time-dependent problems to be considered will be formulated in the interval $[a, b]$ of the real line and in the time interval $[0, T]$. Therefore, the governing differential equation will be satisfied in the domain $[a, b] \times [0, T]$ of the $x - t$ *plane* (Fig. 4.4). For one space dimension, Eq. (4.31) reduces to

$$\frac{\partial c}{\partial t} + \frac{\partial (cv)}{\partial x} = g_S (c, x, t) . \tag{4.35}$$

A solute-transport problem that is well-posed can be stated as adhering to the *Basic Principle for Well posed Problems of Non-diffusive Transport,* defined as follows:

Find the concentration, $c(x, t)$, for $t > 0$, usually up to certain time T, and every $a < x < b$ when values of the concentration are known both initially and at one end of the interval $[a, b]$.

The corresponding mathematical problem is an *initial-boundary-value problem* with *Dirichlet boundary conditions* that can be stated as follows:

Find the function, $c(x, t)$, defined in the rectangle $[a, b] \times [0, T]$, which satisfies the differential equation (4.35) there, together with the initial conditions

$$c(x, 0) = c_0(x), \quad a \le x \le b \tag{4.36}$$

and the boundary conditions

$$c(a, t) = c_\partial(t), \quad 0 \le t \le T. \tag{4.37}$$

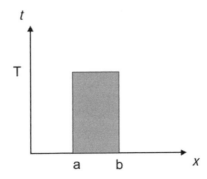

Figure 4.4   Space-time domain for one-dimensional transport problems.

This problem is well-posed whenever $v \neq 0$ everywhere. This condition, $v \neq 0$, is required in order to make sure that any fluid particle that crosses the left boundary of the space interval ($x = a$) will never return to it; if a particle crosses that boundary more than once in the time interval $0 \le t \le T$, then Eq. (4.37) would prescribe the value of c at two different points on the space-time trajectory for the same particle, and the basic principle for well-posed problems of non-diffusive transport would be violated.

The fact that the partial differential equation governing non-diffusive transport reduces to a family of ordinary differential equations, each of them satisfied along the trajectory of each fluid particle, is very useful. Indeed, based on it, we explain next a general method of solution for the boundary-value problem defined by Eqs. (4.35) to (4.37).

A more explicit form of Eq. (4.35) is

$$\frac{\partial c}{\partial t} + v \frac{\partial c}{\partial x} = -c \frac{\partial v}{\partial x} + g_S(c, x, t). \tag{4.38}$$

The *Lagrangian representation* of the concentration, as given by Eq. (1.7), is

$$C(X, t) = c(p(X, t), t). \tag{4.39}$$

Taking the derivative with respect to time of this equation and using Eqs. (4.35), (1.11), and (1.14) it is seen that

$$
\begin{aligned}
\frac{\partial C}{\partial t}(X, t) &= \frac{\partial c}{\partial t}(p(X, t), t) + \frac{\partial c}{\partial x}(p(X, t), t)\frac{\partial p}{\partial t}(X, t)\\
&= \frac{\partial c}{\partial t}(p(X, t), t) + v(p(X, t), t)\frac{\partial c}{\partial x}(p(X, t), t) =\\
&= -c(p(X, t), t)\frac{\partial v}{\partial x}(p(X, t), t) + g_S(c(p(X, t), t), p(X, t), t).
\end{aligned} \tag{4.40}
$$

In conclusion, if we keep fixed the *fluid particle* X, Eq. (4.40) reduces to

$$\frac{dC}{dt}(X, t) = -c(p(X, t), t)\frac{\partial v}{\partial x}(p(X, t), t) + g_S(c(p(X, t), t), p(X, t), t). \tag{4.41}$$

Here we use the *total derivative* because when the fluid particle is kept fixed, the partial and total derivatives coincide. If the *position function*, $p(X, t)$, is known, then Eq. (4.41) constitutes an ordinary differential equation for the concentration along each *fluid-particle trajectory*. Thus, this result exhibits an instance of the general result stated at the beginning of this section: namely, that first-order partial differential equations can be reduced to a family of first-order ordinary differential equations, each satisfied along each characteristic curve. Thereby, the fluid-particle trajectories are shown to be the characteristic curves in the case of non-diffusive solute transport.

As already mentioned, the position function $p(X, t)$ of each particle needs to be known in order to apply Eq. (4.41); however, usually this is not the case, albeit the velocity is indeed known and by the definition of the particle velocity,

$$\frac{dp}{dt}(X, t) = v(p(X, t), t), \text{ when the particle } X \text{ is kept fixed.} \tag{4.42}$$

Equation (4.42) supplies an ordinary differential equation for $p(X, t)$ from which it can be obtained. In conclusion, the general method of solution we referred to above consists of first solving Eq. (4.42) numerically and afterward, when the function $p(X, t)$ is available, integrating Eq. (4.41). At present, ordinary differential equations are easily solved using numerical schemes such as the *Runge-Kutta method*, although simpler procedures are satisfactory in many cases.

As an illustration of this general result, we consider a case when an exact analytical solution can be obtained. To this end, it is assumed that no *solute sources* are present, $g_S(c, \underline{x}, t) = 0$, and that v is a constant. Then Eq. (4.41) reduces to

$$\frac{dC}{dt}(X, t) = 0 \tag{4.43}$$

which can be readily integrated to obtain

$$C(X, t) = \bar{C}(X). \tag{4.44}$$

Here $\bar{C}(X)$ is a function of the fluid-particle location X, exclusively; that is, each fluid particle conserves through time its concentration value.

For this simple case, the solution of the initial-boundary-value problem of Eqs. (4.35) to (4.37) can be obtained as is explained next. Let $x \in [a, b]$, $t \in [0, T]$ and assume that $v > 0$; then for any fluid-particle there are only two possibilities (see subdomains Ω_1 and Ω_2 in Fig. 4.5):

1. It was already on the interval $[a, b]$ at $t = 0$; or

2. It entered the interval $[a, b]$ through its left boundary, $x = a$, at $t \geq 0$.

In the first case, the position of each particle at time $t \geq 0$ is given by

$$x = X + vt. \tag{4.45}$$

Here, X is the position of the fluid particle when $t = 0$. Furthermore, the concentration is

$$c(x, t) = c_0(x - vt). \tag{4.46}$$

In the second case, each fluid-particle can be identified by the time at which it entered the interval $[a, b]$, which necessarily lies between zero and T, and that time will be denoted by t'. Then the position of the particle at any time $t > 0$, such that $t' \leq t \leq T$, is given by

$$x = X + v(t - t'). \tag{4.47}$$

The concentration, on the other hand, of such a particle is given by

$$c(x, t) = c_\partial\left(t - \frac{x - a}{v}\right). \tag{4.48}$$

Observe that given any x and t such that $a \leq x \leq b$ and $0 < t \leq T$, Eq. (4.46) applies when $x \geq a + vt$, while Eq. (4.48) applies when $x < a + vt$.

A further comment is appropriate at this point. The model of the same physical problem, when the system is diffusive and the boundary conditions are Dirichlet, is governed by the differential equation

$$\frac{\partial c}{\partial t} + \frac{\partial(cv)}{\partial x} = \frac{\partial}{\partial x}\left(D\frac{\partial c}{\partial x}\right) + g_s(c, \underline{x}, t) \tag{4.49}$$

and satisfies the same initial conditions. However, the Dirichlet boundary conditions for it are

$$c(a, t) = c_{\partial 1}(t) \quad \text{and} \quad c(b, t) = c_{\partial 2}(t), \quad 0 \leq t \leq T. \tag{4.50}$$

Clearly, the main difference between these boundary conditions and those of Eq. (4.38) is that the latter were supplemented by a condition at the other end point of the space interval $[a, b]$ in which the problem is posed. And this additional boundary condition has to be imposed as long as $D > 0$, independent of how small it may be.

In the theory of partial differential equations, this radical change in behavior is explained as follows. The differential equation of Eq. (4.49) is a *parabolic equation,*

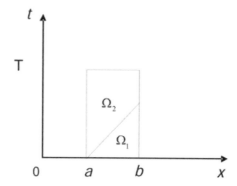

Figure 4.5   Subdomains occupied by particles that entered the problem domain at $t > 0$ and by particles that were already there at $t = 0$.

while that of Eq. (4.35) is only first order (sometimes identified as *hyperbolic*). *Perturbations* of differential equations around certain values of their coefficients are said to be *singular*, when the type of the equation changes whenever the perturbation is different from zero, independent of how small it may be. A conspicuous feature of *singular perturbations of partial differential equations* is that certain boundary conditions that are suitable for the perturbed equation cannot be fulfilled by the unperturbed equation. In this respect, Eq. (4.49) is a *singular perturbation* of the differential equation governing non-diffusive transport, Eq. (4.35), and in this case the boundary condition on the right-hand end of the space interval, which is an adequate boundary condition when $D > 0$, cannot be satisfied when the transport is non-diffusive; i.e., $D = 0$. A very important and well-known example of a singular perturbation of partial differential equations is supplied by the equations governing the flow of a *Newtonian viscous fluid*, which constitute a singular perturbation of the equations that govern *non-viscous fluids*. A phenomenon that can be observed in the solutions of the perturbed equations in such cases is the formation of *boundary layers*; these are narrow strips that accommodate the boundary conditions that cannot be fulfilled by the solutions of the unperturbed equations and whose widths tend to zero as the magnitude of the perturbation tends to zero.

The solution to the problem whose solution appears in Fig. 4.1 is presented in Fig. 4.6 but with the value of D set to zero. The boundary condition on the problem is

$$\frac{\rho_i}{\rho} = c_i = 1, \qquad x = 0$$

and the initial condition is

$$\frac{\rho_i}{\rho} = c_i = 0, \qquad x > 0.$$

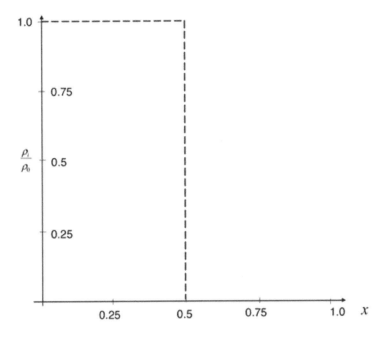

Figure 4.6   Calculated concentrations when there is no diffusion, that is when $D = 0$.

4.8.2 Well-posed problems in several spatial dimensions

The ideas explained in the preceding subsection can easily be extended to multi-dimensional models of non-diffusive transport. The first-order partial differential equation of Eq. (4.31) applies in this case, instead of which Eq. (4.35) can be written as

$$\frac{\partial c}{\partial t} + \underline{v} \cdot \nabla c + (\nabla \cdot \underline{v}) c = g_s (c, \underline{x}, t).\qquad(4.51)$$

We recall that $\underline{v}(\underline{x}, t)$ is assumed to be known; therefore, $\nabla \cdot \underline{v}$ is also known as a function of $\underline{x}$ and t. Thus, when Eq. (4.51) is written using the Lagrange representation of the solute concentration, it becomes [see Eq. (4.40)]

$$\frac{\partial C}{\partial t} (\underline{X}, t) = g_s \left(c \left(\underline{p}(\underline{X}, t), t \right), \underline{p}(\underline{X}, t), t \right) - c \left(\underline{p}(\underline{X}, t), t \right) \nabla \cdot \underline{v} \left(\underline{p}(\underline{X}, t), t \right).$$
$$(4.52)$$

When the fluid particle, $\underline{X}$, is kept fixed, Eq. (4.52) constitutes an ordinary differential equation. At present such an equation can be solved applying numerical methods when the position function $\underline{p}(\underline{X}, t)$ is known.

This function, $\underline{p}(\underline{X}, t)$, can be obtained when the definition of the fluid-particle velocity,

$$\frac{\partial \underline{p}}{\partial t} (\underline{X}, t) = \underline{v} \left(\underline{p}(\underline{X}, t), t \right)\qquad(4.53)$$

is interpreted as a system of ordinary differential equations. Indeed, for a three-dimensional model, a more explicit form of Eq. (4.53) is

$$\frac{\partial p_i}{\partial t}(\underline{X}, t) = v_i\left(\underline{p}(\underline{X}, t), t\right), \quad i = 1, 2, 3. \tag{4.54}$$

Let us now consider time-dependent problems formulated in a domain Ω of the three-dimensional Euclidean space R^3 and in the time interval $[0, T]$. Then the initial and boundary conditions of well-posed problems need to be such that the value of the concentration is prescribed at one and only one point of each characteristic curve (that is, only once on each fluid particle).

4.8.3 Well-posed problems for steady-state models

When the solution sought is time-independent, the governing equation reduces to

$$\nabla \cdot (c\underline{v}) = g_S(c, \underline{x}). \tag{4.55}$$

All the data for this problem must be time-independent in order to be compatible with the steady-state condition, and initial conditions are not prescribed. In particular, the fluid-particle velocity, the source term, and the boundary conditions are generally functions of position but not of time. The characteristic curves are the space trajectories of the fluid particles, and for Dirichlet boundary conditions the concentration value is prescribed at one and only one point of each characteristic curve. For example, when the problem is posed in a domain Ω, one may choose to prescribe the concentration where each fluid particle enters Ω (that is, $\underline{v} \cdot \underline{n} < 0$), in which case it cannot also be prescribed where the fluid-particles exit Ω (that is, $\underline{v} \cdot \underline{n} > 0$).

When diffusion is neglected, the concentration has to satisfy the differential equation

$$v\frac{dc}{dx} = 0 \quad \text{for every } x \text{ such that } 0 < x < 1 \tag{4.56}$$

and the Dirichlet boundary condition

$$c = 1 \text{ at } x = 0. \tag{4.57}$$

The solution of this problem is $c(x) = 1$ for $0 \leq x < 1$.

4.9 SUMMARY

In this chapter we considered the transport of a dissolved species in a free fluid. We discussed the processes of advection, diffusion, and mass generation. We then developed the governing transport equations and the conditions necessary for presenting a well-posed problem for both the transient and steady-state cases. Equilibrium-controlled and kinetic chemical reactions were considered for the mass-generation term. The chapter concluded with a discussion of the governing equations and conditions for defining a well-posed problem for the special case of non-diffusive transport.

EXERCISES

4.1 Derive Eqs. (4.7) to (4.9) from Eq. (4.6).

4.2 The mass of solute contained in a domain fixed in the physical space, $M_S^\Omega(t)$, is given by Eq. (4.11). Show that when the solute transport is conservative and non-diffusive,

$$\frac{dM_S^\Omega}{dt}(t) = - \int_{\partial\Omega} c\underline{v} \cdot \underline{n}\, d\underline{x}. \tag{4.58}$$

4.3 Derive Eqs. (4.16) to (4.20) from Eq. (4.6).

4.4 Consider a pond of cubic shape that contains water in its interior. The lateral walls that limit it are permeable, and the steady-state motion of the water is maintained by suitable means, such as pumping of the water. For maintenance a conservative purifier is added on the top (i.e., the water free surface) in such a manner that its concentration there is kept fixed at the value c_T (T, from top). Furthermore, the diffusion is isotropic with diffusion coefficient equal to D.

Under the assumption that the diffusive flux through the lateral boundaries (or walls) and bottom of the pond vanishes, formulate a well-posed problem suitable for predicting the evolution through time of the solute spatial distribution when an initial state is known;

  a) Show that when c_T is known the rate at which mass of solute is being added can be determined and design a procedure to achieve this goal:

  b) Give an expression, in terms of the concentration of the solute, for the rate at which the total solute mass is being added (notice that such a rate is time-dependent);

  c) Observe, furthermore, that at different points of the free surface the rate at which solute mass is added generally is not the same. Give an expression, as a function of position and time, of such a rate.

4.5 This is similar to Exercise 4.4, except that now the purifier distribution is in steady state. Thus; Under the assumption that the diffusive flux through the lateral boundaries (or walls) and bottom of the pond vanishes, formulate a well-posed boundary-value problem suitable for predicting the steady-state spatial distribution.

  a) Show that when c_T is known, the rate at which the total solute mass of solute is being added can be determined, and design a procedure to achieve this goal;

  b) Give an expression, in terms of the concentration of the solute, for the rate at which the total solute mass is being added (notice that such a rate is time-independent);

  c) Observe, furthermore, that at different points of the free surface the rate at which solute mass is added is not the same. Give an expression, as a function of position, of such a rate.

4.6 Due to sanitary regulations the purifier concentration must remain above a minimum value c_R, specified by the sanitary regulations. Under the assumption that the solute concentration is in steady state and using the results of Exercise 4.5, design a procedure for determining the minimum mass rate at which the purifier must be added at the free surface of the pond.

4.7 A fluid containing a highly toxic substance (for example, a radioactive material) has to be eliminated in order to protect the environment. A procedure, sometimes used, consists of placing the fluid in a deposit and burying it underground. Formulate a boundary-value problem governing the steady-state distribution of the concentration of the toxic substance. For this purpose, assume that:

a) The fluid in the deposit is at rest;
b) The diffusion is isotropic, with diffusion coefficient equal to D;
c) The diffusive flux through the walls of the deposit is proportional to the difference of the concentrations inside and outside the walls (assume that the concentration outside vanishes). Observe that the boundary conditions in this case are of the Robin type.

4.8 As shown in Section 4.6, for first-order irreversible processes the basic differential equation is given by Eq. (4.28). When the parameter λ is independent of time, as is usually the case, by a simple change of variable this equation can be transformed into Eq. (4.6). Indeed, define

$$\widehat{c}\,(\underline{x}, t) \equiv e^{\lambda t} c\,(\underline{x}, t) \tag{4.59}$$

and show that when $c\,(\underline{x}, t)$ satisfies Eq. (4.30), then $\widehat{c}\,(\underline{x}, t)$ fulfills the equation

$$\frac{\partial \widehat{c}}{\partial t} + \nabla \cdot \left(\widehat{c}\,\underline{v} \right) - \nabla \cdot \left(D \nabla \widehat{c} \right) = 0. \tag{4.60}$$

4.9 Consider the non-diffusive and conservative ($g_S = 0$) transport of a solute. Show that the governing differential equation for this process is

$$\frac{\partial c}{\partial t} + \underline{v} \cdot \nabla c = 0. \tag{4.61}$$

Obtain the concentration distribution of the solute, for $t > 0$ and $x > 0$, when:

a) The fluid velocity is a constant;
b) The initial value is $c\,(x, 0) = 1$; and
c) The boundary condition is $c\,(0, t) = e^{-kt}$.

4.10 In Exercise 4.9, assume that in the solute a first-order reaction takes place ($g_S = -\lambda c$), so that the transport is non-conservative. Show that the governing differential equation is

$$\frac{\partial c}{\partial t} + \underline{v} \cdot \nabla c + \lambda c = 0. \tag{4.62}$$

Obtain the concentration distribution of the solute, for $t > 0$ and $x > 0$, when:

a) The fluid velocity is a constant;

b) The initial value is $c(x, 0) = 1$; and

c) The boundary condition is $c(0, t) = e^{-kt}$.
   Discuss the solution for the following three cases: $k < \lambda$, $k = \lambda$, and $k > \lambda$.

REFERENCES

1. Bird, B.R., W.E. Stewart, and E.N. Lightfoot, *Transport Phenomena,* Wiley, New York, 1960.

2. Garabedian, P.R., *Partial Differential Equations*, Wiley, New York, 1964.

3. Grove, D.B. and K.G. Stollenwerk, *Computer Model of One-Dimensional Equilibrium-Controlled Sorption Processes*, U.S. Geological Survey Water-Resources Investigations Report 84-4059, 1984.

4. Zheng, C.H. and G.D. Bennett, *Applied Contaminant Transport Modeling*, Van Nostrand Reinhold, Wiley, New York, 1995.

CHAPTER 5

FLOW OF A FLUID IN A POROUS MEDIUM

The basic mathematical model of fluid flow in a porous medium was introduced in Section 3.2. However, for greater clarity, here we deal with this problem *ab initio*.

5.1 BASIC ASSUMPTIONS OF THE FLOW MODEL

The flow model to be presented in this chapter is based on a set of assumptions that is widely used in subsurface hydrology. They are listed below.

- The porous-medium matrix is saturated with the fluid. As previously explained, this means that the fluid thoroughly fills the pore space of the solid matrix;

- The solid matrix remains at rest throughout the fluid-flow process;

- The solid matrix is elastic. More precisely, the porosity of the matrix is a function of the fluid pressure. Therefore, the pore-level porosity of the matrix may change as time goes by. This, in spite of the fact that the macroscopic velocity of the solid matrix is zero throughout the whole process;

Mathematical Modeling in Science and Engineering: An Axiomatic Approach.
By Ismael Herrera and George F. Pinder Copyright © 2012 John Wiley & Sons, Inc.

- The fluid is compressible. Specifically, the density of the fluid satisfies an equation of state in which the density is a function of the pressure, exclusively;

- The fluid particle velocity fulfills Darcy's law. This is an empirical constitutive equation, which relates the fluid particle velocity to the fluid-pressure distribution;

- The fluid is not subjected to diffusion processes, so that τ equals zero.

The fluid pressure is also referred to as the *pore pressure*, especially in *soil mechanics*. The pressure in the pores pushes on the grain surfaces, exerting a force that tends to increase the pores' volume, just as a balloon inflates under the pressure of the air contained in its interior. In general, the grains move in response to changes in fluid pressure such that the porosity increases as the fluid pressure increases. Concomitantly, a decrease in pore fluid pressure results in a decrease in pore volume leading to *consolidation*. Consolidation can be of practical importance since it may lead to soil deformation at the Earth's surface with associated building instability.

As for the density of the fluid, generally it is a function not only of its pressure, but also of temperature; this implies that the temperature, itself, may be a function of the pressure. Such a function is determined by the kind of thermodynamic processes that the fluid performs during its motion. Two alternative processes are usually assumed: either such a process is isothermal (the temperature does not change) or it is adiabatic (there is no heat exchange). However, the assumption here adopted is that the density is only a function of its pressure.

In addition, the fluid density may be a function of the concentration of dissolved species. Thus, in assuming density to be only a function of pressure, we are implicitly accepting negligible dependence of fluid density on species concentration.

5.2   THE BASIC MODEL FOR THE FLOW OF A FLUID THROUGH A POROUS MEDIUM

We recall the axiomatic method for deriving the mathematical models of continuous systems described in Chapters 1 and 3. The flow system is a two-phase system since it consists of the solid matrix and the fluid contained in its pores. However, the fact that the motion of the solid phase is known, since it is at rest, permits dealing with the fluid phase exclusively and treating the system as a single-phase system. This single phase, in turn, is made of only one component, the fluid. Thus, the family of extensive properties consists of only one extensive property: namely, the *fluid mass*. Recalling that the solid matrix is saturated, it is seen that the mass of the fluid contained in a domain $B(t)$ occupied by the fluid-porous system is given by

$$M_f(t) \equiv \int_{B(t)} \varepsilon(\underline{x}, t) \, \rho(\underline{x}, t) \, dx. \tag{5.1}$$

We are interested in following the motion of fluid bodies, so $B(t)$ will move with the fluid velocity.

When the solid matrix is saturated by the fluid, the porosity, ε, occurring in Eq. (5.1), equals the fraction of the physical space occupied by the fluid. Then $\varepsilon\rho$ is the mass of fluid per unit volume of the physical space occupied by a fluid porous-medium system. This explains Eq. (5.1), which implies that the intensive property associated with the fluid mass is $\varepsilon\left(\underline{x}, t\right)\rho\left(\underline{x}, t\right)$. Therefore, the balance differential equation is given by

$$\frac{\partial\varepsilon\rho}{\partial t} + \nabla\cdot\left(\varepsilon\rho\underline{v}\right) = g \tag{5.2}$$

while the jump condition is

$$\llbracket\varepsilon\rho\left(\underline{v} - \underline{v}_\Sigma\right)\rrbracket\cdot\underline{n} = 0. \tag{5.3}$$

5.3 MODELING THE ELASTICITY AND COMPRESSIBILITY

Under the assumptions of Section 5.1, the product $\varepsilon\rho$ occurring in Eqs. (5.2) and (5.3) is a function of the pressure exclusively. And for the application of such equations, it is necessary to make more explicit the dependency of $\varepsilon\rho$ on the fluid pressure. In particular, the goal of the following development is to express the time derivative of $\varepsilon\rho$, occurring in Eq. (5.2), in terms of the time derivative of the pressure.

To this end, we proceed to decompose the time derivative of $\varepsilon\rho$ into two contributions: one due to the *fluid compressibility* and the other due to the *elasticity of the solid matrix*. Such decomposition follows immediately when the derivative-of-a-product formula is applied:

$$\frac{\partial\varepsilon\rho}{\partial t} = \varepsilon\frac{\partial\rho}{\partial t} + \rho\frac{\partial\varepsilon}{\partial t}. \tag{5.4}$$

Here the term $\frac{\partial\rho}{\partial t}$ yields the contribution of the fluid compressibility, while the term $\frac{\partial\varepsilon}{\partial t}$ yields the contribution of the solid-matrix elasticity as a whole (not the individual grains alone).

5.3.1 Fluid compressibility

In this subsection we make use of the assumption that the fluid satisfies an *equation of state*, which permits expressing the density as a function, $\rho\left(p\right)$, of pressure exclusively. On the other hand, during the process of fluid flow the pressure is a function of position, $\underline{x}$, and time, t. However, the fluid itself is assumed to be homogeneous; that is, it satisfies the same equation of state at every space location and time. Therefore, the density as a function of position and time is given by $\rho = \rho\left(p\left(\underline{x}, t\right)\right)$, which implies that

$$\frac{\partial\rho}{\partial t} = \frac{d\rho}{dp}\frac{\partial p}{\partial t} \equiv \beta\rho\frac{\partial p}{\partial t}. \tag{5.5}$$

This equation defines the parameter β, which is known as the *fluid compressibility*, and more explicitly, it is defined by

$$\beta \equiv \frac{1}{\rho}\frac{d\rho}{dp} = -\frac{1}{V}\frac{dV}{dp}. \tag{5.6}$$

Here V stands for the *specific volume* of the fluid, which is defined by

$$V \equiv \rho^{-1}. \tag{5.7}$$

In words: The specific volume is the volume per unit mass.

In Eq. (5.6) we have used the relation

$$\frac{1}{\rho}\frac{d\rho}{dp} = \frac{d\,\ell n\rho}{dp} = -\frac{d\,\ell n\rho^{-1}}{dp} = -\frac{d\,\ell nV}{dp} = -\frac{1}{V}\frac{dV}{dp}. \tag{5.8}$$

5.3.2 Pore compressibility

To understand the processes that produce and determine the compressibility of the solid matrix, we present a very succinct stress analysis. Let p_{tot} be the *total pressure* (force per unit area) attributable to the whole solid-fluid system.[1] Part of the system is supported by the solid matrix and another part by the fluid (Fig. 5.1). The notation p_{ef} (from *effective pressure*) is used for the support provided by the solid matrix. Then

$$p_{tot} = p_{ef} + p. \tag{5.9}$$

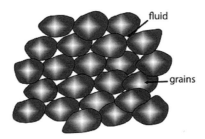

Figure 5.1   Porous medium showing matrix and fluid.

Observe that the magnitude of p_{tot} depends on the conditions of the environment surrounding the porous medium-fluid system. For example, if such a system constitutes the soil on which a building is positioned, p_{tot} will certainly change if the building is removed. In the analysis that follows, it is assumed that the conditions of the environment surrounding the porous fluid system do not change, so that the total pressure does not change during the time considered in the analysis. Some problems considered in soil mechanics and foundation engineering consist of studying

---

[1] Here an isotropic stress state is assumed.

the modifications in the fluid pressure distribution produced by a change in p_{tot} due, for example, to the construction of civil works such as buildings. A similar analysis, although not included here, can be applied to such a problem.

When the assumption of a time-independent p_{tot} is adopted, any change in the fluid pressure is accompanied by a change in the effective pressure. Because of Eq. (5.9), this observation implies that

$$\Delta p_{ef} + \Delta p = \Delta p_{tot} = 0. \tag{5.10}$$

Here, but only here, the symbol Δ stands for *increment* or *change*. Notice that when the pore pressure increases, the effective pressure decreases and concomitantly, the pores expand.

The following notation will be used next:

$$\rho_s \equiv \text{density of the solid material}$$
$$V_s \equiv \text{specific volume of the solid} = \rho_s^{-1}$$

and

$$\rho_{tot} \equiv \text{density of the solid matrix}$$
$$V_{tot} \equiv \text{specific volume of the solid matrix} = \rho_{tot}^{-1}.$$

Furthermore, the compressibility of the solid matrix β_{tot} and the compressibility of the solid grains β_s are given, respectively, by

$$\beta_{tot} \equiv -\frac{1}{V_{tot}} \frac{dV_{tot}}{dp_{ef}} \text{ and } \beta_s \equiv -\frac{1}{V_s} \frac{dV_s}{dp_{ef}} = \rho_s \frac{d\rho_s}{dp_{ef}}.$$

We observe that

$$\rho_{tot} = (1 - \varepsilon) \rho_s \text{ and } V_{tot} = \frac{1}{(1 - \varepsilon) \rho_s}. \tag{5.11}$$

We start our analysis with the identity

$$\varepsilon = 1 - \frac{(1 - \varepsilon) \rho_s}{\rho_s} = 1 - \frac{V_s}{V_{tot}}. \tag{5.12}$$

Taking the derivative of this relationship and multiplying and dividing the second term by V_s, one obtains

$$\frac{d\varepsilon}{dp_{ef}} = \frac{V_s}{V_{tot}} \frac{1}{V_{tot}} \frac{dV_{tot}}{dp_{ef}} - \frac{V_s}{V_{tot}} \frac{1}{V_s} \frac{dV_s}{dp_{ef}} \tag{5.13}$$

which can be written using Eq. (5.12) as

$$\frac{d\varepsilon}{dp_{ef}} = (\beta_s - \beta_{tot}) \frac{V_s}{V_{tot}} = (\beta_s - \beta_{tot}) (1 - \varepsilon). \tag{5.14}$$

Therefore,

$$\frac{d\varepsilon}{dp} = -\frac{d\varepsilon}{dp_{ef}} = (\beta_{tot} - \beta_s) (1 - \varepsilon). \tag{5.15}$$

Hence,

$$\frac{\partial \varepsilon}{\partial t} = \frac{d\varepsilon}{dp}\frac{\partial p}{\partial t} = (\beta_{tot} - \beta_s)(1 - \varepsilon)\frac{\partial p}{\partial t}. \tag{5.16}$$

Usually, $\beta_{tot} \gg \beta_s$ and β_s is neglected. Then

$$\frac{\partial \varepsilon}{\partial t} = \beta_{tot}(1 - \varepsilon)\frac{\partial p}{\partial t}. \tag{5.17}$$

The condition $\beta_{tot} \gg \beta_s$ is fulfilled when the change of volume of the pores is much larger than the change in volume of the solid material that forms the solid matrix.

5.3.3 The storage coefficient

Making use of Eqs. (5.4), (5.5), and (5.16), we get

$$\frac{\partial \varepsilon \rho}{\partial t} = \rho\{\varepsilon\beta + (1 - \varepsilon)(\beta_{tot} - \beta_s)\}\frac{\partial p}{\partial t}. \tag{5.18}$$

The *specific storage coefficient*, S_S, is defined to be

$$S_S \equiv \rho\hat{g}\{\varepsilon\beta + (1 - \varepsilon)(\beta_{tot} - \beta_s)\}. \tag{5.19}$$

Here $\hat{g}$ is the *gravitational acceleration*. The balance differential equation, Eq. (5.2), after multiplying by $\hat{g}$, can now be written as

$$S_S\frac{\partial p}{\partial t} + \hat{g}\nabla \cdot (\varepsilon\rho\underline{v}) = -\rho\hat{g}q. \tag{5.20}$$

In this equation the rate of external supply of mass, g, has been written in a form that is usual in groundwater hydrology:

$$g = -\rho q \tag{5.21}$$

where q represents the volume rate of withdrawal, per unit volume, of the fluid.

5.4 DARCY'S LAW

This law is an empirical constitutive equation that relates the fluid velocity and the fluid pressure spatial distribution. It was established in the nineteenth century by the French engineer Henri Darcy for one-dimensional saturated flow through sand, but since then has been generalized to consider more complicated flow regimes. In particular, in a generalized form, it is also now used to describe multiphase flow in anisotropic porous media. Here we consider the case when the fluid has only one phase; multiphase flows will be discussed in Chapters 7 and 8.

Generally, when the solid matrix is anisotropic, the porous medium may have preferred directions for the flow of fluids. For that general case, *Darcy's law* for a single-phase fluid is given by the equation

$$\underline{U} \equiv \varepsilon\underline{v} = -\frac{1}{\mu}\underline{\underline{k}} \cdot (\nabla p - \rho\hat{g}). \tag{5.22}$$

Here

$\hat{g}$ is the acceleration due to gravity (a vector)

μ is the *dynamic viscosity* of the fluid

$\underline{\underline{k}}$ is the *intrinsic permeability tensor*

$\underline{U}$ is the Darcy velocity.

The intrinsic permeability tensor, $\underline{\underline{k}}$, is a positive definite and symmetric matrix.

We note the similarity between this law and *Fourier's (flow of heat)* and *Fick's (flow of solute mass) laws*. However, in Darcy's law the force of gravity plays a special role, something that does not happen in the other two laws. It should also be pointed out that in flows in which Darcy's law applies, the fluid pressure is always continuous. This is necessary because otherwise the gradient of the pressure would be of infinite magnitude, and so also would be the fluid velocities.

As already stated, Eq. (5.22) applies in the general case when the porous matrix may be anisotropic. In the particular case when the solid matrix is isotropic there are no preferred directions for the flow due to the solid matrix, and the intrinsic permeability tensor, $\underline{\underline{k}}$, has the form

$$\underline{\underline{k}} = k\underline{\underline{I}}. \tag{5.23}$$

Here the *intrinsic permeability* k is a scalar. When Eq. (5.23) holds, the intrinsic permeability tensor is positive definite if and only if $k > 0$.

Given any point $\underline{x}$ in the physical space, let $z(\underline{x})$ be its elevation with respect to a given reference level. Then the acceleration due to gravity is

$$\hat{g} = -\hat{g}\nabla z \tag{5.24}$$

where, as earlier, $\hat{g}$ is the magnitude of the gravitational acceleration. Using this equation, Eq. (5.22) becomes

$$\underline{U} = -\frac{1}{\mu}\underline{\underline{k}} \cdot (\nabla p + \rho\hat{g}\nabla z) \tag{5.25}$$

or, using indicial notation,

$$U_i = -\frac{k_{ij}}{\mu}\left(\frac{\partial p}{\partial x_j} + \rho\hat{g}\frac{\partial z}{\partial x_j}\right); \qquad i = 1, 2, 3. \tag{5.26}$$

Here,

$$\underline{U} = \begin{pmatrix} U_1 \\ U_2 \\ U_3 \end{pmatrix}. \tag{5.27}$$

When the solid matrix is isotropic, Eq. (5.25) reduces to

$$\underline{U} = -\frac{1}{\mu}k(\nabla p + \rho\hat{g}\nabla z). \tag{5.28}$$

5.5 PIEZOMETRIC LEVEL

When modeling flow through porous media, especially in the study of groundwater hydrology, the concept of *piezometric head* or *piezometric level* is very useful. To introduce it, we start by defining an auxiliary function:

$$H\left(p, z\right) \equiv \frac{1}{\hat{g}} \int_{p_0}^{p} \frac{d\xi}{\rho\left(\xi\right)} + z. \tag{5.29}$$

We observe that when the fluid is incompressible, $\rho\left(\xi\right)$ is a constant independent of ξ and Eq. (5.29) becomes

$$H\left(p, z\right) \equiv \frac{p - p_0}{\rho\hat{g}} + z. \tag{5.30}$$

Then, at any time t and any point $\underline{x}$ of a saturated porous medium, we define the *piezometric level*, $h\left(\underline{x}, t\right)$, by

$$h\left(\underline{x}, t\right) \equiv H\left(p\left(\underline{x}, t\right), z\left(\underline{x}\right)\right). \tag{5.31}$$

Or, more explicitly,

$$h\left(\underline{x}, t\right) \equiv \frac{1}{\hat{g}} \int_{p_0}^{p(\underline{x}, t)} \frac{d\xi}{\rho\left(\xi\right)} + z\left(\underline{x}\right). \tag{5.32}$$

When the fluid is incompressible, this yields

$$h\left(\underline{x}, t\right) \equiv \frac{p\left(\underline{x}, t\right) - p_0}{\rho\hat{g}} + z\left(\underline{x}\right). \tag{5.33}$$

Water has very little compressibility under normal conditions. Due to this fact, Eq. (5.33) is used extensively in groundwater hydrology.

We notice, for later use, that using Eq. (5.32) we can write

$$\frac{\partial p}{\partial t} = \rho\hat{g}\frac{\partial h}{\partial t}. \tag{5.34}$$

Another important property of the piezometric head, which in turn motivates its extensive use, is, again using Eq. (5.32),

$$\nabla p + \rho\hat{g}\nabla z = \rho\hat{g}\nabla h. \tag{5.35}$$

Thus Eq. (5.25) can be written as

$$\underline{U} = -\frac{\rho\hat{g}}{\mu}\underline{\underline{k}} \cdot \nabla h. \tag{5.36}$$

The *hydraulic conductivity tensor* is defined to be

$$\underline{\underline{K}} \equiv \frac{\rho\hat{g}}{\mu}\underline{\underline{k}}. \tag{5.37}$$

Then, Eq. (5.36) can be written

$$\underline{U} = -\underline{\underline{K}} \cdot \nabla h. \tag{5.38}$$

When the solid matrix is isotropic,

$$\underline{\underline{K}} = \frac{\rho \hat{g}}{\mu} k \underline{\underline{I}} = K \underline{\underline{I}}. \tag{5.39}$$

The *hydraulic conductivity* is defined to be

$$K \equiv \frac{\rho \hat{g}}{\mu} k. \tag{5.40}$$

Using it, in the isotropic case, one has

$$\underline{U} = -K \nabla h. \tag{5.41}$$

In subsurface hydrology Eqs. (5.38) and (5.41) are commonly used to express the Darcy velocity, while the forms used in the oil industry are closer to Eqs. (5.25) and (5.28).

The expressions for Darcy velocity given by Eqs.(5.36) and (5.38) are actually very general, since in them the tensors $\underline{\underline{k}}$ and $\underline{\underline{K}}$ may be anisotropic. As noted, these tensors are always symmetric and positive definite, and only in the limiting case for which the medium is impermeable may they vanish. It can be shown, applying spectral theory to the matrices $\underline{\underline{k}}$ and $\underline{\underline{K}}$, that the positiveness condition is tantamount to requiring that the fluid always flows from regions where the piezometric head is larger to those where it is smaller. This is similar to the condition, in solute transport, where diffusive flow must go from regions where the concentration is larger to those where it is smaller.

When studying subsurface flow, the tensors $\underline{\underline{k}}$ and $\underline{\underline{K}}$ are defined by the properties of the materials that constitute the strata of the subsoil. Frequently such material has been generated through processes in which gravity has played a special role, as is the case for sedimentary strata. As a consequence, frequently the *intrinsic permeability* and the *hydraulic conductivity tensors* have the vertical direction as an *axis of symmetry*. Given this constraint, the permeability in all horizontal directions is the same but is usually different from that of the vertical direction.

Any matrix, $\underline{\underline{M}}$, with a symmetry axis can be expressed as

$$\underline{\underline{M}} = M_T \underline{\underline{I}} + (M_L - M_T) \underline{e} \otimes \underline{e} \tag{5.42}$$

where $\underline{e}$ and M_L are a unit vector and the *proper value* (eigenvalue) in the direction of the symmetry axis, respectively, while M_T is the proper value in the transverse directions (that is, in any direction orthogonal to the symmetry axis). As we have done throughout this book, here we use $\underline{\underline{I}}$ for the identity matrix.

Applying Eq. (5.42) to the case when the intrinsic permeability and the hydraulic conductivity tensors have the vertical direction as an *axis of symmetry*, we write, as is customary,

$$\underline{\underline{k}} = k_H \underline{\underline{I}} + (k_V - k_H) \underline{e}_z \otimes \underline{e}_z \text{ and } \underline{\underline{K}} = K_H \underline{\underline{I}} + (K_V - K_H) \underline{e}_z \otimes \underline{e}_z. \tag{5.43}$$

Here $\underline{e} \otimes \underline{e}$ is a matrix given by

$$\underline{e} \otimes \underline{e} \equiv \begin{pmatrix} e_1 e_1 & e_1 e_2 & e_1 e_3 \\ e_2 e_1 & e_2 e_2 & e_2 e_3 \\ e_3 e_1 & e_3 e_2 & e_3 e_3 \end{pmatrix}.$$

In this equation, k_H and k_V are the horizontal and vertical intrinsic permeabilities, respectively; similarly, K_H and K_V are the horizontal and vertical hydraulic conductivities, respectively. As for $\underline{e}_z$, it is a unit vector in the vertical direction.

5.6   GENERAL EQUATION GOVERNING FLOW THROUGH A POROUS MEDIUM

Combination of Eq. (5.34) with Eq. (5.20) yields

$$S_S \frac{\partial h}{\partial t} + \rho^{-1} \nabla \cdot (\rho \underline{U}) = -q \tag{5.44}$$

that is,

$$S_S \frac{\partial h}{\partial t} + \nabla \cdot \underline{U} + \underline{U} \cdot \nabla (\ell n\, \rho) = -q \tag{5.45}$$

When the fluid is slightly compressible and Darcy's velocity is moderate, the term $\underline{U} \cdot \nabla \ell n\, \rho \ll 1$ and can be neglected. Then, we write

$$S_S \frac{\partial h}{\partial t} + \nabla \cdot \underline{U} = -q. \tag{5.46}$$

Equation (5.46) is often used to derive models describing the flow of a single-phase fluid through a porous medium, as we shall do here.

When Darcy's law, as given by Eq. (5.38), is incorporated in Eq. (5.46), one gets

$$S_S \frac{\partial h}{\partial t} - \nabla \cdot \left(\underline{\underline{K}} \cdot \nabla h \right) = -q. \tag{5.47}$$

Here, as earlier, q is the volumetric rate of extraction of the fluid from the porous medium: that is the volume of fluid withdrawn per unit time, per unit volume of the porous system.

This is the basic differential equation that is used extensively in applications, particularly in subsurface hydrology. Equation (5.47) and those that will be derived from it are linear differential equations when its coefficients, the specific storage and the hydraulic conductivity, are independent of h. Actually, these coefficients depend to some extent on h, but in many applications that dependence is neglected and the equations are treated as linear equations, as we shall do in what follows.

5.6.1 Special forms of the governing differential equation

The governing differential equation, as given by Eq. (5.47), is very general and there are several special forms that should be highlighted. When the porous medium is isotropic, Eq. (5.39) applies and Eq. (5.47) reduces to

$$S_S \frac{\partial h}{\partial t} - \nabla \cdot (K \nabla h) = -q. \tag{5.48}$$

If, furthermore, the solid matrix is homogeneous, its properties are independent of position and one can write

$$S_S \frac{\partial h}{\partial t} - K \nabla^2 h = -q. \tag{5.49}$$

Or, using the notation $\nabla^2 \equiv \Delta$,

$$S_S \frac{\partial h}{\partial t} - K \Delta h = -q. \tag{5.50}$$

This in turn can be written as

$$\frac{\partial h}{\alpha \partial t} - \Delta h = -q/K \tag{5.51}$$

where $\alpha \equiv K/S_S$. This form of the equation has interest from the perspective of dimensional and similarity analysis, since by a change of variables, in which $\alpha t \rightarrow t$, it can be transformed into the *heat equation*, when $q = 0$; then:

$$\frac{\partial h}{\partial t} - \Delta h = 0. \tag{5.52}$$

Another interesting special case is when the fluid-porous system is incompressible (that is, the compressibilities of both the porous matrix and the fluid are negligible). Such a case corresponds to the case when $S_S = 0$, so that Eqs. (5.47), (5.48), and (5.50) reduce to

$$\nabla \cdot \left(\underline{\underline{K}} \cdot \nabla h \right) = q \tag{5.53}$$

and

$$\nabla \cdot (K \nabla h) = q \tag{5.54}$$

and

$$\nabla^2 h = q/K \tag{5.55}$$

respectively.

On the other hand, for steady-state problems the time derivatives in Eqs. (5.47) to (5.52) vanish and, therefore, Eqs. (5.53) to (5.55) also apply to steady-state problems. Recall that Eq. (5.54) assumes isotropy of the solid matrix, while Eq. (5.55) assumes, in addition, that the matrix properties are independent of position (that is, that the porous matrix is homogeneous). On the other hand, Eq. (5.53) is the more general equation describing time-independent problems with no restrictions on hydraulic conductivity.

5.7   APPLICATIONS OF THE JUMP CONDITIONS

For relatively simple applications such as those corresponding to problems of ground-water supply, discontinuities that have to be considered in the flow models are due mainly to the lithology of the aquifers, which gives rise to sharp changes in the properties of the solid matrix. The axiomatic formulation of the basic equations of the continuous model permits developing this subject in a very systematic manner.

When two strata of different geologic origin are in contact, generally they carry different porosity and permeability, and these properties jump across the interface, separating them. For the case of flow through a porous medium, the jump conditions are given by Eq. (5.3), which when the Darcy velocity is incorporated reads

$$[\![\rho\,(\underline{U} - \varepsilon\underline{v}_\Sigma)]\!] \cdot \underline{n} = 0. \tag{5.56}$$

We recall that $\underline{v}_\Sigma$ here is the velocity of the surface where the discontinuity occurs. When Σ is the interface separating two geological strata that are fixed in the physical space, $\underline{v}_\Sigma$ vanishes identically. Then Eq. (5.56) reduces to

$$[\![\rho\underline{U}]\!] \cdot \underline{n} = 0. \tag{5.57}$$

As pointed out in Section 5.4, Darcy's law implies that the fluid pressure is continuous and therefore so is the fluid density. Hence

$$[\![\underline{U}]\!] \cdot \underline{n} = 0. \tag{5.58}$$

In summary, the normal component of the Darcy velocity is continuous across the interface separating two strata possessing different properties. However, in general, the *fluid particle velocity* is *discontinuous* there, due to a change in the porosity. Indeed, Eq. (5.58) implies that

$$\varepsilon_+\underline{v}_+ \cdot \underline{n} = \varepsilon_-\underline{v}_- \cdot \underline{n}. \tag{5.59}$$

Hence, $\underline{v}_+ \cdot \underline{n} \neq \underline{v}_- \cdot \underline{n}$ whenever $\varepsilon_+ \neq \varepsilon_-$. Here the subscripts plus and minus have been used to distinguish adjacent strata, one from the other.

5.8   WELL-POSED PROBLEMS

The class of problems that are well posed for the various models is determined by the type of the governing differential equations. In the case of flow through a porous medium, two types of differential equations will be encountered: *parabolic* and *elliptic*. For flow through a porous medium, the most general governing equation is given by Eq. (5.47), which is a parabolic partial differential equation, as long as $S_S > 0$, because the hydraulic conductivity tensor, $\underline{K}$, is always a positive definite matrix. However, when $S_S = 0$ and for steady-state models, the governing differential equation reduces to a partial differential equation of elliptic type. We recall that these two types of equations also occurred when studying solute transport by a free fluid as described in Section 4.4 and, consequently, the discussion that follows is very similar to that presented there. However, in spite of the similarities between the mathematical models governing these two kinds of systems there are significant differences between the underlying physics that should be stressed.

Although the governing equation for solute transport by a free fluid, as given by Eq. (4.6), is of parabolic type, when that equation is compared with Eq. (5.47) it exhibits some relevant differences that reflect the different physical principles at work. A very important one, which has significant implications for its numerical treatment and the properties of the resulting numerical solutions, is the fact that Eq. (4.6) contains an advection (or convection) term, $c\underline{v}$, that is absent in Eq. (5.47). Due to this fact, the differential operator involving the space coordinates associated with Eq. (5.47) is a symmetric one, while it is non-symmetric for Eq. (4.6). We also observe that the coefficient of the second-order derivative is the scalar D, in Eq. (4.6), while it is a matrix $\underline{\underline{K}}$ for Eq. (5.47).

5.8.1 Steady-state models

To start with, we consider well-posed problems for steady-state models for which the governing differential equations are of elliptic type. In this case, well-posed problems are *boundary-value problems* that seek to obtain a function $h(\underline{x})$ that satisfies Eq. (5.53) or Eq. (5.54) or Eq. (5.55) in a domain Ω of the physical space, and also satisfies suitable boundary conditions, which are discussed next, on its boundary, $\partial\Omega$.

For Eq. (5.54) the most general class of boundary conditions are a type known as the *generalized Robin boundary conditions*; these have the following form:

$$\alpha\underline{U}(\underline{x})\cdot\underline{n} + \beta(\underline{x})h(\underline{x}) = \gamma(\underline{x}), \quad \forall\underline{x}\in\partial\Omega. \tag{5.60}$$

Here again, as in Eq. (4.22) on page 70, we take α and β so that $\alpha^2 + \beta^2 = 1$. In view of the fact that $\underline{U}$ is the Darcy velocity, the term $\underline{U}(\underline{x})\cdot\underline{n}$ represents the volumetric flow per unit area that flows out from the domain Ω through its boundary, $\partial\Omega$.

When Darcy's law is incorporated, Eq. (5.60) becomes

$$-\alpha\underline{n}\cdot\underline{\underline{K}}\nabla h(\underline{x}) + \beta(\underline{x})h(\underline{x}) = \gamma(\underline{x}), \quad \forall\underline{x}\in\partial\Omega. \tag{5.61}$$

In particular, when the porous medium is isotropic, then

$$-\alpha K\frac{\partial h}{\partial n} + \beta(\underline{x})h(\underline{x}) = \gamma(\underline{x}), \quad \forall\underline{x}\in\partial\Omega \tag{5.62}$$

which is a Robin boundary condition of the type discussed in Section 4.5 on page 69. We observe that Eq. (5.60) is indeed a more general form of Robin boundary condition than that given by Eq. (4.22).

5.8.1.1 Problems with the prescribed piezometric head
There are two extreme cases of the generalized Robin boundary condition introduced above. One corresponds to a *Dirichlet problem* ($\alpha = 0$); then Eq. (5.60) adopts the form

$$h(\underline{x}) = h_\partial(\underline{x}), \quad \forall\underline{x}\in\partial\Omega. \tag{5.63}$$

Given a porous medium and a fluid (which in subsurface hydrology would be an aquifer saturated with water) in steady-state conditions, the practical problem consists of predicting the piezometric head distribution in the interior of the porous medium when the head distribution is known only on the outer boundary of the porous medium. Obviously, when a mathematical and computational model is not available, if a distribution of the heads in Ω were required, it would be necessary to measure head values in a sufficiently dense mesh that covers the whole domain Ω and then interpolate these values to generate a head surface. On the other hand, if a reliable model is used, it would only be necessary to carry out such measurements on the boundary $\partial\Omega$, which implies considerable savings of resources and time.

In applications, when an aquifer is limited by a lake or a river, the head equals the level of the surface water body, so the head is known beforehand. In such cases it is convenient and appropriate to develop steady-state mathematical models based on boundary-value problems in which the known head levels on the boundary are prescribed.

5.8.1.2 Problems with the prescribed volumetric flow   The other extreme case of the generalized *Robin boundary condition* is a *generalized Neumann problem* ($\beta = 0$). Then Eq. (5.60) reduces to

$$\underline{U}(\underline{x}) \cdot \underline{n} = \gamma(\underline{x}), \quad \forall \underline{x} \in \partial\Omega. \tag{5.64}$$

In this case, $\underline{U}(\underline{x}) \cdot \underline{n}$ is prescribed; when it is positive it describes the volumetric outflow and describes the volumetric inflow when it is negative, per unit time per unit area of the domain boundary. Such boundary conditions, when expressed in terms of the piezometric head as in Eq. (5.61), are

$$\underline{n} \cdot \underline{\underline{K}}\nabla h(\underline{x}) = \gamma(\underline{x}), \quad \forall \underline{x} \in \partial\Omega \tag{5.65}$$

and for an isotropic porous medium, Eq. (5.62),

$$K\frac{\partial h}{\partial n} = \gamma(\underline{x}), \quad \forall \underline{x} \in \partial\Omega. \tag{5.66}$$

Models for which the outflow (or inflow) is prescribed can only be applied when there are available measurements, or estimates, of the volume that enters or leaves the domain where the porous medium is located. In subsurface hydrology, sophisticated procedures have been developed [27] for estimating the recharge of aquifers, and also techniques for distributing such recharge along their boundaries.

Aquifers are usually formed by geological strata whose horizontal dimensions are much larger than their vertical dimensions; and permeable strata are frequently limited vertically by relatively impermeable strata. In such cases the aquifer consisting of the permeable strata is said to be a *confined aquifer*. The boundary condition that is suitable for this kind of aquifer is the condition of no flow at the interface between the permeable and impermeable strata. In a more general form applicable to any point on the domain boundary, this boundary condition can be written as

$$\underline{U}(\underline{x}) \cdot \underline{n} = 0. \tag{5.67}$$

When use is made of Darcy's law, Eq. (5.67) becomes

$$\underline{n} \cdot \underline{K} \nabla h \left(\underline{x} \right) = 0. \tag{5.68}$$

For an isotropic porous medium this reduces to

$$\frac{\partial h}{\partial n} = 0 \quad \text{at the interface.} \tag{5.69}$$

5.8.2 Time-dependent problems

The governing differential equations for time-dependent problems, when $S_S > 0$, are parabolic. Then well-posed problems are *initial-boundary-value problems*, which seek a function $h \left(\underline{x}, t \right)$ that satisfies Eq. (5.47) in a domain Ω, together with suitable boundary conditions defined on its boundary over a specified time interval. Such boundary conditions may be of any of the types that are suitable for steady-state problems; these were discussed in Section 5.8.1.2. Furthermore, the function $h \left(\underline{x}, t \right)$ is required to satisfy additionally the initial conditions

$$h \left(\underline{x}, 0 \right) = h_0 \left(\underline{x} \right), \forall \underline{x} \in \Omega. \tag{5.70}$$

Here $h_0 \left(\underline{x} \right)$ is a prescribed function that is known beforehand.

Well-posed problems for time-dependent problems in the case when the aquifer is incompressible (that is, when $S_S = 0$) exhibit some special features that deserve attention. In such a case, the governing equations may be any one of Eqs. (5.53) to (5.55). These are partial differential equations of elliptic type and the well-posed problems are *boundary-value problems* for each time, t. As explained before, such problems seek to obtain a function $h \left(\underline{x}, t \right)$ that satisfies one of Eqs. (5.53) to (5.55), in the domain Ω of the physical space, in combination with boundary conditions prescribed on its boundary, $\partial \Omega$. These boundary conditions, which can be of any of the types already discussed, generally will depend on time, so the solution of such elliptic problems will depend on the parameter t. We observe that no *initial conditions* can be satisfied when $S_S = 0$.

Fluids, rocks, and soils that exist in nature are more or less compressible, so the concept of an *incompressible aquifer* must be understood as an approximation. Therefore, for time-dependent problems the solutions that are obtained using Eq. (5.53), or those derived from it, are approximations to those obtained using Eq. (5.47). In particular, if the boundary conditions and the extraction and recharge are kept fixed through time, the solution obtained using $S_S > 0$ approaches the solution obtained with $S_S = 0$, as $S_S \to 0$, except at an interval of time whose length goes to zero as $S_S \to 0$. We may conclude that perturbations of S_S around $S_S = 0$ are similar to *singular perturbations* except that in this case the boundary layer is not in space but instead is in time. This phenomenon is explained in Appendix C.

5.9 MODELS WITH A REDUCED NUMBER OF SPATIAL DIMENSIONS

In all the discussions presented up to now, the physical space has been modeled as a three-dimensional (3-D) Euclidean space. However, in some problems of engineering and science it is useful to apply two-dimensional or even one-dimensional models. This may be justified by reasons that depend on the problem considered. For example, in groundwater hydrology the horizontal dimensions of aquifers are frequently much larger than their thickness and, when studying them, the variations of parameters such as the piezometric head in the vertical direction are so small that they can be neglected.

Of course, in order to use such simplified models in a reliable manner it is necessary to have information about their range of applicability, which can be acquired by theoretical or empirical means. Frequently, the theoretical analysis of the errors introduced by models with a reduced number of dimensions is so complicated that it is not practical to carry it out, and then the only means for establishing the range of applicability are empirical.

However, there are a few reduced-dimension models that are based on very complete, yet relatively uncomplicated theoretical foundations. Next we describe one such example whose analysis is also useful for introducing and illustrating in a natural manner some of the ideas and concepts that are basic for that class of models.

5.9.1 Theoretical derivation of a 2-D model for a confined aquifer

Consider the confined aquifer of Fig. (5.2). The following assumptions are adopted:

1. Its thickness is uniform;

2. The aquifer is vertically homogeneous (that is, the properties of the material that constitutes the solid matrix do not depend on its vertical coordinate);

3. The vertical direction is a symmetry axis for the hydraulic conductivity tensor (that is, any vector in the vertical direction is a proper vector of the hydraulic conductivity matrix); and

4. The stratum that constitutes the aquifer is limited by two low-permeable strata, lying above and below (that is, the aquifer is a confined aquifer).

Taking a system of Cartesian coordinates in which every point of the space is given by $\underline{x} \equiv (x_1, x_2, x_3)$, we define

$$z \equiv x_3. \tag{5.71}$$

We assume that the aquifer occupies that portion of the space for which $0 \le z \le b$. Then, applying Eq. (5.67), it is seen that

$$U_3(0) = U_3(b) = 0. \tag{5.72}$$

Now we write Eq. (5.46) in the form

$$S_S \frac{\partial h}{\partial t} + \sum_{\alpha=1}^{2} \frac{\partial U_\alpha}{\partial x_\alpha} + \frac{\partial U_3}{\partial z} = -q. \tag{5.73}$$

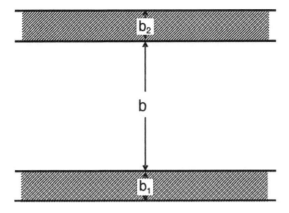

Figure 5.2   Diagrammatic sketch of confined aquifer.

Now integrate this equation from $z = 0$ to $z = b$, and use Eq. (5.72), to obtain,

$$S_s \frac{\partial}{\partial t} \int_0^b h \, dz + \sum_{\alpha=1}^2 \int_0^b \frac{\partial U_\alpha}{\partial x_\alpha} dz = - \int_0^b q \, dz. \qquad (5.74)$$

The *storage coefficient* and the *total volume withdrawal per unit area* are given by

$$S \equiv \int_0^b S_S \, dz = b \ \text{ and } \ Q \equiv \int_0^b q dz \qquad (5.75)$$

respectively. Using these definitions, Eq. (5.74) reads

$$S \frac{\partial \bar{h}}{\partial t} + b \sum_{\alpha=1}^2 \frac{\partial \bar{U}_\alpha}{\partial x_\alpha} = - Q \qquad (5.76)$$

where the averages of the piezometric head and of the velocities are given by:

$$\bar{h} \equiv \frac{1}{b} \int_0^b h dz$$
$$\bar{U}_\alpha \equiv \frac{1}{b} \int_0^b U_\alpha dz. \qquad (5.77)$$

Using Darcy's law and the assumed symmetry of the hydraulic conductivity tensor, it is seen that

$$\bar{U}_\alpha = -K_H \frac{\partial \bar{h}}{\partial x_\alpha}. \qquad (5.78)$$

Here K_H is the *horizontal conductivity*. Therefore, Eq. (5.76) can be transformed into

$$S \frac{\partial \bar{h}}{\partial t} - \sum_{\alpha=1}^2 \frac{\partial}{\partial x_\alpha} \left(T \frac{\partial \bar{h}}{\partial x_\alpha} \right) = Q. \qquad (5.79)$$

Here

$$T \equiv bK_H \tag{5.80}$$

and is called the *transmissivity*.

We observe that Eq. (5.79) is an *exact* equation for the *head average* and therefore when it is subject to suitable initial and boundary conditions, it makes it possible in principle, to obtain the *exact* values of the *head average*. When the *head* variations across the aquifer thickness are small, its vertical average constitutes a good approximation of its value at any point across the aquifer.

5.9.2 Leaky aquitard method

Hantush [14] discussed the possibility that the material above the upper boundary of the aquifer could be permeable, albeit the permeability would be low (aquitard). Under these circumstances significant quantities of water could enter the aquifer through vertical leakage. Figure 5.3 illustrates the type of system to be envisioned. The analysis of this system is taken largely from Pinder and Celia [27]. In this system, the aquifer is bounded from above by a low-permeability layer (layer A) relative to the aquifer. While of low permeability, this layer is nevertheless capable of providing water to the aquifer via *vertical leakage*. The base of the aquifer is also bounded by a low-permeability layer (layer B). Below the aquitard is an almost impermeable layer. It is assumed with good justification that the water in the low permeability layers moves only vertically. Above layer A is an unpumped aquifer which maintains a constant head during the pumping test. The aquifer is assumed to be of uniform thickness, to have infinite areal extent, and to be homogeneous.

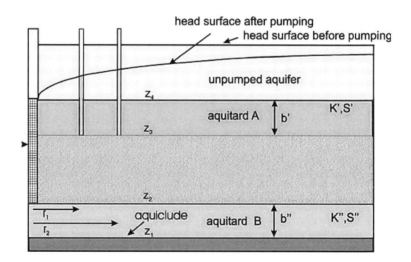

Figure 5.3   Definition sketch of aquifer system addressed by Hantush leaky aquitard solution.

The groundwater flow equation that describes this system is a variant on Eq. (5.79); that is,

$$\frac{\partial^2}{\partial r^2} h\left(r,t\right) + \frac{1}{r}\frac{\partial}{\partial r} h\left(r,t\right) - \frac{S}{T}\frac{\partial}{\partial t}\left(h\left(r,t\right)\right) + \frac{K'}{T}\frac{\partial h_A}{\partial z} - \frac{K''}{T}\frac{\partial h_B}{\partial z} = 0 \quad (5.81)$$

where K' and S' and K'' and S' are the hydraulic conductivities and specific storage coefficients in layer A and layer B, respectively. The one-dimensional form of the groundwater flow equation can be used to describe the transient head distribution in the aquitards, that is,

$$\frac{S'}{K'b'}\frac{\partial}{\partial t}\left(h_A\left(z,t\right)\right) - \frac{\partial}{\partial z}\left[\frac{\partial}{\partial z}\left(h_A\left(z,t\right)\right)\right] = 0 \quad (5.82)$$

and

$$\frac{S''}{K''b''}\frac{\partial}{\partial t}\left(h_B\left(z,t\right)\right) - \frac{\partial}{\partial z}\left[\frac{\partial}{\partial z}\left(h_B\left(z,t\right)\right)\right] = 0. \quad (5.83)$$

The solution of this set of equations requires specification of initial and boundary values for each of the *state variables* h, h_A, and h_B. Let us define the *drawdown* $s = H - h$ and $s_n = H_n - h_n$, where $n = A, B$, and H, and H_n are the initial head values in the system. The boundary conditions and initial conditions can now be stated as follows: For the upper aquitard the initial condition is

$$s_A\left(r,z,0\right) = 0. \quad (5.84)$$

At the top of the upper aquitard the boundary condition is

$$s_A\left(r,z_4,t\right) = 0 \quad (5.85)$$

and at the bottom it is

$$s_A\left(r,z_3,t\right) = s\left(r,t\right) \quad (5.86)$$

where z_1, z_2, z_3, and z_4 are defined in Fig. 5.3.

The initial condition for the aquifer is given as

$$s\left(r,0\right) = 0 \quad (5.87)$$

and the boundary condition at $r \to \infty$ is

$$\lim_{r\to\infty} s\left(r,t\right) = 0. \quad (5.88)$$

The boundary condition at the infinitely small well bore is given by

$$\lim_{r\to 0}\left(r\frac{\partial s}{\partial r}\right) = -\frac{Q}{2\pi T}. \quad (5.89)$$

The physical meaning of this relationship is seen by cross-multiplying r and T. One now sees that flow to the well is balanced by the flow through the well perimeter, which has a circumference of $2\pi r$. In the lower aquitard, the initial condition is

$$s_B(r, z, 0) = 0. \tag{5.90}$$

On the top of the bottom aquitard the boundary condition (that is, $z = z_2$) is

$$s_B(r, z_2, t) = s(r, t) \tag{5.91}$$

and on the bottom ($z = z_1$) it is

$$\frac{\partial s_B(r, z_1, t)}{\partial z} = 0.$$

A *small time* solution to this system of equations was suggested by Hantush and is discussed by Batu [2]. The necessary conditions for the application of this solution are

$$\frac{b'S'}{K'} \geq 10r \tag{5.92}$$

and

$$\frac{b''S''}{K''} \geq 10r. \tag{5.93}$$

The solution form is

$$s(r, t) = \frac{Q}{4\pi T} H(u, \beta) \tag{5.94}$$

where

$$H(u, \beta) = \int_u^\infty \frac{e^{-y}}{y} \text{erfc} \left\{ \frac{\beta u^{\frac{1}{2}}}{[y(y - u)]^{\frac{1}{2}}} \right\} dy \tag{5.95}$$

$$u = \frac{r^2 S}{4Tt} \tag{5.96}$$

and

$$\beta = \frac{r}{4} \left[\left(\frac{K'S'}{b'TS} \right)^{\frac{1}{2}} + \left(\frac{K''S''}{b''TS} \right)^{\frac{1}{2}} \right]. \tag{5.97}$$

5.9.3 The integrodifferential equations approach

As already mentioned, the applicability of the Hantush solutions just presented is restricted to small periods of time. This kind of analytical solution is very useful in subsurface hydrology when carrying out single-well analysis, such as in pump test interpretations, in which the small times restriction is frequently fulfilled. Note, however, that Neuman and Witherspoon [26] developed a solution that is not subject to such a constraint. On the other hand, transient leakage calculations are also needed for regional aquifer system analysis in which groundwater behavior in confining units is a significant component of the total water budget. In such cases one has to resort to computational codes based on numerical models [23]. A thorough approach based on integrodifferential equations, which has served as a basis for constructing regional numerical models of leaky-aquifer systems [23], was introduced and developed by Herrera et al. ([15]-[21]). A summary of this approach is contained in reference [23] where additional contributions to this subject are included [7], [10], [12], [28].

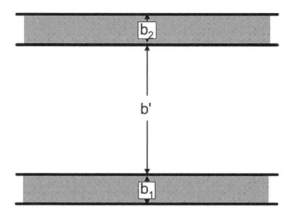

Figure 5.4   The aquitard with thickness b' is bounded from above and below by permeable aquifers of thickness b_2 and b_1.

Here we explain the integrodifferential equations approach for the system of two aquifers separated by an aquitard as shown in Fig. 5.4 (we draw mainly from references [18], [20] and [23]). For simplicity, here we only discuss the equations that govern aquifer 1, but similar equations apply to aquifer 2. In terms of *drawdown* and in Cartesian coordinates, for aquifer 1 Eq. (5.79) implies that

$$\frac{1}{\alpha}\frac{\partial s}{\partial t} - \frac{\partial^2 s}{\partial x^2} - \frac{\partial^2 s}{\partial y^2} = -Q_L \tag{5.98}$$

with

$$Q_L \equiv -\frac{K'}{T}\left(\frac{\partial s'}{\partial z}\right)_{z=0}. \tag{5.99}$$

Here $s'(x, y, z, t)$ satisfies the conditions

$$\frac{\partial^2 s'}{\partial z^2} = \frac{1}{\alpha'}\frac{\partial s'}{\partial t}, \quad 0 < t, \quad 0 < z < b' \tag{5.100}$$

together with

$$s'(x, y, z, 0) = 0, \quad s'(x, y, 0, t) = s(x, y, t), \quad s'(x, y, b', t) = s_2(x, y, t). \tag{5.101}$$

Here, as in what follows, primes are used to distinguish aquitard properties and the subscript 2 refers to aquifer 2. When $s(x, y, t)$ and $s_2(x, y, t)$ are given, the problem defined by Eqs. (5.100) and (5.101) is well-posed; its solution $s'(x, y, z, t)$ can be expressed by means of Duhamel's integrals (see [18] and [23]), in the form

$$s'(x, y, z, t) = \int_0^t \frac{\partial s}{\partial t}(x, y, t - \tau)\, u(z, \tau)\, d\tau + \int_0^t \frac{\partial s_2}{\partial t}(x, y, t - \tau)\, v(z, \tau)\, d\tau.$$

(5.102)

Here, $u(z, t)$ and $v(z, t)$ are auxiliary functions that fulfill Eq. (5.100) subjected to

$$\left. \begin{array}{ll} u(0, t) = 1, & v(0, t) = 0 \\ u(b', t) = 0, & v(0, t) = 1 \end{array} \right\}, \quad t > 0$$
$$u(z, 0) = 0, \quad v(z, 0) = 0, \quad 0 < z < b'.$$

(5.103)

The solution of these initial-boundary-value problems is given in reference [18]. Applying Eq. (5.99) and evaluating $(\partial s'/\partial z)_{z=0}$ by means of the representation of Eq. (5.102) yields (see [17] for details)

$$\begin{aligned} Q_L(x, y, t) &= \frac{K'}{Tb'} \int_0^t \frac{\partial s_1}{\partial t}(x, y, t - \tau)\, f\left(\alpha'\tau/b'^2\right) d\tau \\ &- \frac{K'}{Tb'} \int_0^t \frac{\partial s_2}{\partial t}(x, y, t - \tau)\, h\left(\alpha'\tau/b'^2\right) d\tau. \end{aligned}$$

(5.104)

Here

$$f\left(\alpha'\tau/b'^2\right) \equiv b'\frac{\partial u}{\partial z}(0, \tau) \quad \text{and} \quad h\left(\alpha'\tau/b'^2\right) \equiv -b'\frac{\partial v}{\partial z}(0, \tau).$$

(5.105)

The governing integrodifferential equations are obtained when Eqs. (1.7) and (1.8) are used in Eq. (1.1):

$$\begin{aligned} \frac{1}{\alpha}\frac{\partial s}{\partial t} &= \frac{\partial^2 s}{\partial x^2} + \frac{\partial^2 s}{\partial y^2} \\ &- \frac{K'}{Tb'} \int_0^t \frac{\partial s}{\partial t}(x, y, t - \tau)\, f\left(\alpha'\tau/b'^2\right) d\tau \\ &- \frac{K'}{Tb'} \int_0^t \frac{\partial s_2}{\partial t}(x, y, t - \tau)\, h\left(\alpha'\tau/b'^2\right) d\tau. \end{aligned}$$

(5.106)

In general, the latter equation is coupled with a similar equation for aquifer 2. However, when the piezometric head of aquifer 2 remains unperturbed through time (that is, $s_2 \equiv 0$), Eq. (5.106) reduces to

$$\frac{1}{\alpha}\frac{\partial s}{\partial t} = \frac{\partial^2 s}{\partial x^2} + \frac{\partial^2 s}{\partial y^2} - \frac{K'}{Tb'} \int_0^t \frac{\partial s}{\partial t}(x, y, t - \tau)\, f\left(\alpha'\tau/b'^2\right) d\tau$$

(5.107)

and can be solved separately when it is supplemented by suitable initial and boundary conditions.

For a homogeneous aquifer system the function f has two equivalent expressions:

$$f\left(\alpha't/b'^2\right) = 1 + 2\sum_{n=1}^{\infty} e^{-n^2\pi^2\alpha't/b'^2} \quad \text{and}$$

$$f\left(\alpha't/b'^2\right) = \frac{1}{(\pi\alpha't/b'^2)^{1/2}}\left(1 + 2\sum_{n=1}^{\infty} e^{-n^2b'^2/\alpha't}\right). \quad (5.108)$$

We observe that when the time t is sufficiently small, one has

$$f\left(\alpha't/b'^2\right) \approx \frac{1}{(\pi\alpha't/b'^2)^{1/2}}. \quad (5.109)$$

Another manner of expressing f that is relevant is

$$f\left(\alpha't/b'^2\right) = 1 + g\left(\alpha't/b'^2\right) \quad \text{where} \quad g\left(\alpha't/b'^2\right) \equiv 2\sum_{n=1}^{\infty} e^{-n^2\pi^2\alpha't/b'^2}. \quad (5.110)$$

When Eq. (5.110) is used, Eq. (5.107) takes the form

$$\frac{1}{\alpha}\frac{\partial s}{\partial t} = \frac{\partial^2 s}{\partial x^2} + \frac{\partial^2 s}{\partial y^2} - \frac{K'}{Tb'}\left\{s + \int_0^t \frac{\partial s}{\partial t}\left(x, y, t-\tau\right) g\left(\alpha'\tau/b'^2\right) d\tau\right\}. \quad (5.111)$$

Here, as in what follows, it is assumed that the initial condition for the drawdown of aquifer 1 is $s\left(x, y, 0\right) \equiv 0$. As for the function h, it is given by

$$h\left(\alpha't/b'^2\right) = 1 + 2\sum_{n=1}^{\infty} (-1)^n e^{-n^2\pi^2\alpha't/b'^2}. \quad (5.112)$$

The functions f and h are illustrated in Fig. 5.5. For the efficient numerical use of the integrodifferential equations, some special procedures of integration have been developed (see [20] and [23]).

It has been shown (see [18]) that Hantush's *small times* solution is the exact solution of the integrodifferential equation of Eq. (5.107) with f approximated by Eq. (5.109). Furthermore, Hantush also supplied a *long times* solution [14] that has been shown to be the exact solution of Eq. (5.111) with g given by

$$g\left(\alpha'\tau/b'^2\right) = \alpha'/3b'^2\delta\left(\tau\right). \quad (5.113)$$

where $\delta\left(\tau\right)$ is Dirac's delta function. The meaning of this approximation is that the full water storage of the aquitard is released instantly when the drawdown occurs. This is because

$$\alpha'/b'^2 \int_0^{\infty} g\left(\alpha'\tau/b'^2\right) d\tau = 1/3. \quad (5.114)$$

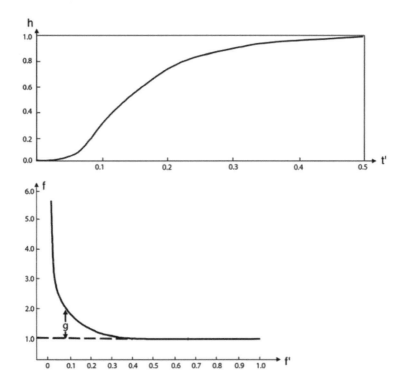

Figure 5.5   Top panel is the influence function h. The lower panel shows the memory functions f and g.

5.9.4   Other 2-D aquifer models

When deriving approximate models, one must distinguish two steps:

1. The model formulation; and

2. The error evaluation.

By definition, an approximate model must predict the behavior of the system except for a *small* error, and therefore approximate models are proposed with the expectation that they will produce small errors within a suitable range of applications. In many procedures for deriving approximate models these two steps are so closely related that it is difficult to separate one from the other. However, there are many instances in which the formulation is to a large extent independent of the error evaluation. This is the case generally when the proposed approximate model is suggested not by mathematical analysis but, rather, by practical experience.

A procedure that yields a wider class of 2-D aquifer models than that presented in Section 5.9.1 is based on the application, in a two-dimensional physical space, of the axiomatic method for deriving models of continuous systems. Using it we will obtain 2-D models that can be applied not only to confined aquifers, but to unconfined aquifers as well. They are based on the following assumptions:

1. The aquifer is vertically homogeneous;

2. The vertical direction is a symmetry axis for the *hydraulic conductivity tensor*;

3. The aquifer is confined at its bottom by an impermeable layer, but may be either confined or unconfined at its top;

4. Each vertical section of the aquifer is in hydrostatic equilibrium; that is, the piezometric head, h, is independent of the altitude z; and

5. The fluid is incompressible.

Under these assumptions, the piezometric head is only a function of two coordinates, x_1 and x_2, and we will write $\underline{x} \equiv (x_1, x_2)$ during derivation of this model. Furthermore, the Darcy velocity is horizontal, independent of z, and is given by

$$\varepsilon \underline{v} = \underline{U}(\underline{x}, t) = -K_H \nabla h(\underline{x}, t). \tag{5.115}$$

Therefore, the fluid-particle velocities are also horizontal and independent of z.

In order to apply the axiomatic approach of Sections 1.1 to 1.3 in two dimensions, we consider *two-dimensional bodies*. Any one such body occupies at any time a domain $\bar{B}(t)$ of the 2-D plane that moves with the fluid-particle velocity $\underline{v}$. With each two-dimensional $\bar{B}(t)$ we associate a 3-D body, $B(t)$ (Fig. 5.6); the latter 3-D body is a cylinder whose base is $\bar{B}(t)$ and height is $b(\underline{x}, t)$. Here $b(\underline{x}, t)$ is a function that specifies the height of the cylinder at $\underline{x} \in \bar{B}(t)$ and time t. Therefore, the cylinder $B(t)$ is characterized by the condition that every one of its points $(\underline{x}, z) \in B(t)$ fulfills $\underline{x} \in \bar{B}(t)$ together with $0 \leq z \leq b(\underline{x}, t)$. Hence, using

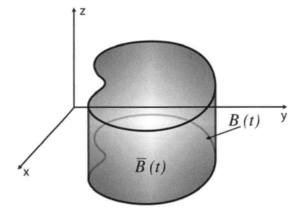

Figure 5.6 Definition of a cylinder.

Eq. (5.1) it is seen that the mass of fluid contained in such a cylinder is given by

$$\bar{M}_f(t) \equiv \int_{\bar{B}(t)} \left(\int_0^b \rho \varepsilon(\underline{x}, t)\, dz \right) dx = \rho \int_{\bar{B}(t)} b(\underline{x}, t)\, \varepsilon(\underline{x}, t)\, dx. \qquad (5.116)$$

Here, use has been made of the vertical homogeneity of the aquifer assumption and ρ, the density of fluid, is a constant since the fluid is incompressible and not a function of either temperature or species concentration. Thus, with each two-dimensional body $\bar{B}(t)$ we associate the mass given by Eq. (5.116), as an area integral over $\bar{B}(t)$, and the desired 2-D model will be derived by the axiomatic approach assuming mass as the sole extensive property. Furthermore, the associated intensive property is the corresponding integrand, in Eq. (5.116); that is,

$$\psi(\underline{x}, t) = \rho b(\underline{x}, t)\, \varepsilon(\underline{x}, z, t). \qquad (5.117)$$

In view of Eq. (1.25), the global mass balance equation is

$$\frac{d\bar{M}_f}{dt}(t) = -\int_{\bar{B}(t)} \rho b(\underline{x}, t)\, q(\underline{x}, t)\, dx = -\int_{\bar{B}(t)} \rho Q(\underline{x}, t)\, dx \qquad (5.118)$$

where

$$Q(\underline{x}, t) \equiv b(\underline{x}, t)\, q(\underline{x}, t) \qquad (5.119)$$

is, as noted earlier, the total volume withdrawal per unit area per unit time. By Eq. (1.32) the mass-balance differential equation is

$$\frac{\partial b\varepsilon}{\partial t} + \nabla \cdot (b\varepsilon \underline{v}) = b\frac{\partial \varepsilon}{\partial t} + \varepsilon \frac{\partial b}{\partial t} + \nabla \cdot (b\underline{U}) = -Q. \qquad (5.120)$$

When use is made of the results of Section 5.3, this equation becomes

$$\varepsilon \frac{\partial b}{\partial t} + S_S \frac{\partial h}{\partial t} - \sum_{\alpha=1}^{2} \frac{\partial}{\partial x_\alpha} \left(bK_H \frac{\partial h}{\partial x_\alpha} \right) = -Q \qquad (5.121)$$

or

$$\varepsilon \frac{\partial b}{\partial t} + S_S \frac{\partial h}{\partial t} - \nabla \cdot (bK_H \nabla h) = -Q. \qquad (5.122)$$

Now we make two applications of Eq. (5.122). The first is to confined aquifers; in that case, b the aquifer thickness is time independent. Hence $\partial b / \partial t = 0$ and Eq. (5.122) reduces to

$$S\frac{\partial h}{\partial t} - \nabla \cdot (T\nabla h) = -Q. \qquad (5.123)$$

So we have recovered Eq. (5.79) using a different approach.

On the other hand, when the aquifer is a free aquifer, $b = h$, and Eq. (5.122) becomes

$$(S + \varepsilon)\frac{\partial h}{\partial t} - \nabla \cdot (K_H h \nabla h) = -Q. \qquad (5.124)$$

This is a non-linear equation well known in the literature; however, in its non-linear form it has a restricted range of applicability [11]. When the piezometric head is close to a fixed value $b(\underline{x})$, independent of time, Eq. (5.124) can be linearized and it then becomes

$$(S + \varepsilon) \frac{\partial h}{\partial t} - \nabla \cdot (T \nabla h) = -Q. \tag{5.125}$$

This equation is frequently applied in regional studies to treat free-surface aquifers.

5.10 SUMMARY

In this chapter we considered the development of equations that describe the flow of a single-phase fluid in a porous medium. After stating the basic assumptions to be employed in the equation development, we provided the fundamental form of the fluid mass-balance equation. The resulting equation requires closure relationships, and these were provided in the form of fluid and porous medium compressibility equations and a simplified version of the momentum balance equation known as Darcy's law. Introduction of the closure relationships into the fluid-balance equation resulted in a final equation for which the well-posedness conditions were next considered. In subsequent sections, simplified forms of the equation were discussed. Expression of the governing equation when the number of spatial coordinate directions are reduced was discussed next, including methodology to approximately recoup the physical impact on the solution of the dimension eliminated.

EXERCISES

5.1 Explore the physical implications of the assumptions stated in Section 5.1 and comment on them. Then:

  a) Give an expression for the volume of the fluid contained in a body that occupies the domain $B(t)$ of the physical space;

  b) At first sight, the assumption that the solid matrix is elastic and yet at rest may seem contradictory. How is it possible that the pore volume changes in time, whereas the solid particles do not move?

  c) Show that when there is only one component in a phase, any diffusive process can be accounted for by modifying the particle velocity.
    Hint. This can be derived from basic principles. Use Eq. (1.27), with $g_\Sigma = 0$, and observe that

$$\frac{d}{dt} \int_{B(t)} \psi \, dx - \int_{\partial B(t)} \underline{\tau} \cdot \underline{n} \, dx = \int_{B(t)} \left\{ \frac{\partial \psi}{\partial t} + \nabla \cdot (\psi \underline{v} - \underline{\tau}) \right\} dx. \tag{5.126}$$

Give, if possible, an expression for modification of the particle velocity.

5.2   Equation (5.4) can be written

$$\frac{\partial \varepsilon \rho}{\partial t} = \varepsilon \frac{\partial \rho}{\partial t} + \rho \frac{\partial \varepsilon}{\partial t} = \left(\varepsilon \frac{d\rho}{dp} + \rho \frac{d\varepsilon}{dp} \right) \frac{\partial p}{\partial t}. \tag{5.127}$$

This equation constitutes an *elegant* manner of making explicit and separating the influence of the fluid properties from those of the solid matrix. Here, we call this equation elegant because (1) it is *clear*; (2) it is *general*, since it does not assume any specific properties of the fluid, nor of the porous medium; and (3)it is *simple* since it was derived from the application of very elementary knowledge–namely, the derivative-of-a-product formula. It can be applied whenever $d\rho/dp$ and $d\varepsilon/dp$ are known. We observe that the latter knowledge is equivalent to knowing both ρ and ε as functions of the pressure. In Sections 5.3.1 and 5.3.2, additional assumptions were introduced to evaluate $d\rho/dp$ and $d\varepsilon/dp$. Therefore, the applicability of the definition of the storage coefficient given in Section 5.3.3, Eq. (5.19), is restricted by such assumptions. As a first exercise on this topic, using the more basic relation of Eq. (5.4), give a revised definition of S_S in terms of $d\rho/dp$ and $d\varepsilon/dp$ in such a manner that Eq.(5.20) holds. Then, by introducing your new definition you have widened the range of applicability of Eq. (5.20).

5.3   Assume that the fluid satisfies the equation of an ideal gas, Eq. (2.76). Then, obtain the density of the gas and its derivative, as functions of the pressure, for the following two cases:

    a) When the process is isothermal, and

    b) When the process is adiabatic.

5.4   In Section 5.3.2, a relatively simple analysis (maybe too simplified) was carried out that permitted us to express the derivative of the porosity in the form given in Eq. (5.1). Sometimes, it is not possible to carry out such theoretical analyses and it is better to resort to more pragmatic approaches, such as field or laboratory experiments. Design a black-box experiment to evaluate $d\varepsilon/dp$.

5.5   Show that for an isotropic medium, Darcy's law implies that in the flow of a fluid in a porous medium, the fluid velocity is proportional to the force per unit volume acting on the fluid phase.

5.6   Take p_0 in Eq. (5.30) equal to the atmospheric pressure. Show that when the fluid is at rest, $h(\underline{x}, t)$ is uniform and equal to the height at which the fluid rises.

5.7   Generally, when the solid matrix is anisotropic, the intrinsic permeability tensor, $\underline{\underline{k}}$, may be any symmetric and positive (actually, non-negative) matrix. Show that due to these properties of $\underline{\underline{k}}$, there are always three principal directions of the intrinsic permeability tensor, and within the range of possibilities there are two extreme cases: one of them the case for which there are three different proper values, and the other the case when all the proper values are equal. Then:

    a) Show that in the latter case the medium is isotropic and $\underline{\underline{k}}$ is given by
        Eq. (5.3):

b) Discuss the intermediate case, for which two of the proper values are equal but the third is different. Show that the intrinsic permeability tensor is isotropic in the plane perpendicular to the proper vector associated with the third proper value.

c) An explicit expression for $\underline{k}$, analogous to Eq. (5.23), is

$$\underline{k} = k_I \underline{I} + (k_V - k_I)\underline{\underline{T}}.\qquad(5.128)$$

Here the matrix $\underline{\underline{T}}$ is given by

$$\underline{\underline{T}} \equiv (T_{ij}) \text{ with } T_{ij} \equiv e_i e_j, \quad i,j = 1,2,3.\qquad(5.129)$$

Furthermore, $\underline{e} = (e_1, e_2, e_3)$ is the proper vector associated with the third proper value.

5.8 Write the general equation of flow, given by Eq. (5.47), for each of the three forms of the intrinsic permeability tensor discussed above. When writing these equations, take the coordinate axes in the direction of three orthogonal proper vectors. Furthermore, assume that the system is homogeneous (i.e., all properties are independent of position).

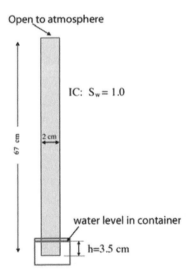

Figure 5.7   Sand-packed, water saturated column open to the atmosphere on top and submerged in water at the bottom.

5.9 In Fig. 5.7 a sand-packed, water-saturated column is placed in a beaker in which the water level is 3.5 cm higher than the base of the sand column. A screen is located in the bottom of the column to prevent sand from escaping. The top of the column is open to the atmosphere. Assume that you are going to model this system using the two-phase air-water flow equations. Specify the boundary conditions you would use to describe this problem assuming that you are working in a two-dimensional $r - z$ coordinate system. You will need boundary conditions for all four sides of the column as pictured.

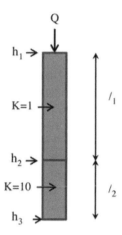

Figure 5.8   Column saturated with water and flowing at steady state; h is the total head and K is the hydraulic conductivity.

5.10   Consider the information in Fig. 5.8. This is a column saturated with water and flowing at steady state; h is the total head and K is the hydraulic conductivity.
   a) Is the water flowing top to bottom or bottom to top, and why?
   b) Is water flowing faster through the $K = 1$ filled portion of the column or through the $K = 10$ portion?
   c) Assume the following values: $l_1 = 10$, $l_2 = 5$, $h_1 = 6$, $h_3 = 3$, calculate the value at h_2.

   Hint: Remember conservation of fluid volume.

5.11   Fig. 5.10.c represents a plan view of a well field. The dots are wells, and the numbers to the right and below the dots are locations in x and y. The numbers to the left and above the dots are water-level elevations.
   a) Calculate the value of the head at the location of the square in Fig. 5.10.c.

   Hint: First write the equation for a plane, then determine the three coefficients defining the plain, and finally calculate the value of the head at the required point.
   Also calculate the head gradient at the location of the square.
   b) If the hydraulic conductivity is 5, what is the magnitude and approximate direction of the specific discharge (i.e., $\mathbf{q}$)?

   Hint: You will need to differentiate the equation of the plane you derived above.

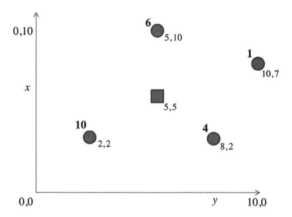

c) If the porosity is 0.5, what is the average groundwater velocity $\mathbf{v}$ (pore velocity)?

REFERENCES

1. Anderson, M.P. and W.W. Woessner. *Applied Groundwater Modeling: Simulation of Flow and Advective Transport*, Academic Press, San Diego, CA, 1991.

2. Batu, V., *Aquifer Hydraulics*, Wiley, New York, 1998.

3. Bear, J., *Dynamics of Fluids in Porous Media*, Elsevier, New York, 1972.

4. Bear, J., *Hydraulics of Groundwater*, McGraw-Hill, New York, 1979.

5. Chen, B. and I. Herrera, I. Numerical treatment of leaky aquifers in the short time range, *Water Resour. Res.*, 18(3), 557-562, 1982

6. Cheng, A.H.-D., *Multilayered aquifer systems*, Marcel Dekker, New York, 2000.

7. Cooley, R.L., *A Modular Finite-Element Model (MODFE) for Areal and Axi-symmetric Ground-water Flow Problems,* Part 2: Derivation of Finite-Element Equations and Comparisons with Analytical Solutions-Theory and Applications: U.S. Geological Survey Techniques of Water-Resources Investigations, book 6, chap. A4, p. 108, 1992.

8. Courant, R. and D. Hilbert, *Methods of Mathematical Physics*, Vol. II: Partial Differential Equations, Wiley-Interscience, New York, 1962.

9. Dake, L.P., *Fundamentals of Reservoir Engineering*, Developments in Petroleum Science, Vol. 8, Elsevier, Amsterdam, 1978.

10. De Marsily, G., E. Ledoux, A. Levassor, D. Poitrinal, and A. Salem, Modeling of large multilayer aquifer systems:theory and applications, *Hydrol.*, 36, 1-34, 1978.

11. De Marsily, G., *Quantitative Hydrogeology, Groundwater Hydrology for Engineers,* Academic Press, New York, 1986.

12. Gambolati, G., F. Sartoreto, and F. Uliana, A conjugate gradient finite element model of for large multiaquifer systems, *Water Resour. Res.*, 22, 1003-1015, 1986.

13. Fanchi, J.R., *Principles of Applied Reservoir Simulation*, 2nd ed. (with CD-Rom), Gulf Professional Publishing, Elsevier, New York, 2001.

14. Hantush, M.S., Modification of the theory of leaky aquifers, *J. Geophys. Res.*, Vol. 65(11), 3713-3725, 1960.

15. Hennart, J.P., R. Yates, and I. Herrera, Extension of the integrodifferential equations approach to inhomogeneous multiaquifer systems, *Water Resour. Res.*, 17(4), 1044-1050, 1981.

16. Herrera, I. and V. Figueroa , A correspondence principle for the theory of leaky aquifers, *Water Resour. Res.*, 5(4), 900-904, 1969.

17. Herrera, I., Theory of multiple leaky aquifers, *Water Resour. Res.*, 6(1), 185-193, 1970.

18. Herrera, I. and L. Rodarte, Integrodifferential equations for systems of leaky aquifers and applications: 1. The nature of approximate theories, *Water Resour. Res.*, 9(4), 995-1005, 1973.

19. Herrera, I., Integrodifferential equations for systems of leaky aquifers and applications: 2. Error analysis of approximate theories, *Water Resour. Res.*, 10(4), 811-820, 1974.

20. Herrera, I. and R. Yates, Integrodifferential equations for systems of leaky aquifers: 3. A numerical method of unlimited applicability, *Water Resour. Res.,* 13(4), 725-732, 1977.

21. Herrera, I., J.P. Hennart, and R. Yates, A critical discussion of numerical models for multiaquifer systems, *Adv. Water Resour.*, 3(4), 159-163, 1980

22. Hutter, K. and K. Jöhnk, *Continuum Methods of Physical Modeling*, Springer-Verlag, Berlin, 2004.

23. Leake, S.A., P. Leahy, and A.S. Navoy, *Documentation of a computer program to simulate transient leakage from confining units using the modular finite-difference ground-water flow model*, U.S. Geological Survey Open-File Report 94-59, 1994.

24. Lee, T.C., *Applied Mathematics in Hydrogeology*, Lewis Publishers, Boca Raton, FL,1999.

25. National Research Council, *Ground Water Models, Scientific and Regulatory Applications*, National Academy Press, Washington, DC, 1990.

26. Neuman, S.P. and P.A. Witherspoon, Theory of flow in a confined two-aquifer system, *Water Resourc. Res.*, 5, 803-816, 1969.

27. Pinder, G.F. and M.A. Celia, *Subsurface Hydrology*, Wiley, Hoboken, NJ, 2006.

28. Premchitt, J., A technique using integrodifferential equations for model simulation of multiaquifer systems, *Water Resour. Res.*, 17, 162-168, 1981.

29. Spitz, K. and J. Moreno, *A Practical Guide to Groundwater and Solute Transport Modeling*, Wiley, New York, 1996.

30. Zheng, C.H. and G.D. Bennett, *Applied Contaminant Transport Modeling*, Van Nostrand Reinhold, Wiley, New York, 1995.

CHAPTER 6

SOLUTE TRANSPORT IN A POROUS MEDIUM

In many respects the discussion that follows is similar to that presented in Chapter 4, which is natural since both the present chapter and Chapter 4 refer to transport of solutes by a fluid. A basic assumption implicit in the transport models of this section is that the fluid motion is known beforehand; in particular, the fluid velocity is provided as a datum. This requires either measuring such velocities or deriving them through an application of a flow model such as that presented in Chapter 5. However, such discussions also differ in several aspects. As in the previous chapter, it will be assumed that the fluid is located in the pores of a porous medium and that such a medium is *saturated;* that is, that the fluid fills completely the pores. Thus, although the medium is saturated, the fluid does not fill the physical space since part of it is occupied by the *solid matrix.*

The basic mathematical model for the kind of systems studied here was introduced in Chapter 3. We recall that the mass of solute, $M_S(t)$, contained in a body of a porous system, which is made up of the solid matrix and the fluid, is given by the integral

$$M_S(t) = \int_{B(t)} \varepsilon(\underline{x}, t) \, c(\underline{x}, t) \, dx. \tag{6.1}$$

Mathematical Modeling in Science and Engineering: An Axiomatic Approach.
By Ismael Herrera and George F. Pinder Copyright © 2012 John Wiley & Sons, Inc.

Here $c(\underline{x}, t)$ is the concentration of the solute, that is, the mass of solute per unit volume of the fluid, and $\varepsilon(\underline{x}, t)$ is the porosity.

On the other hand, the global balance equation is

$$\frac{dM_S}{dt}(t) = \int_{B(t)} g_S(\underline{x}, t)\, dx + \int_{\partial B(t)} \underline{\tau}_S(\underline{x}, t) \cdot \underline{n}(\underline{x}, t)\, dx \qquad (6.2)$$

where $g_S(\underline{x}, t)$ represents the solute mass sources and $\underline{\tau}_S(\underline{x}, t)$ is the solute mass flux, which as we shall see is due mainly to the diffusion-dispersion processes. Now applying the axiomatic formulation of continuous models, the basic mathematical model of Section 3.4 is recovered. It consists of the governing differential equation,

$$\frac{\partial \varepsilon c}{\partial t} + \nabla \cdot (\varepsilon c \underline{v}) = g_S + \nabla \cdot \underline{\tau}_S \qquad (6.3)$$

together with the jump conditions

$$[\![\varepsilon c(\underline{v} - \underline{v}_\Sigma) - \underline{\tau}_S]\!] \cdot \underline{n} = 0. \qquad (6.4)$$

The first of these equations must be fulfilled at every point occupied by the solid-matrix-fluid system; except on any surface, Σ, where a discontinuity of the variables of the model occurs.

6.1 TRANSPORT PROCESSES

As in the case of solute transport by a free fluid, these processes are advection, mass generation (non-conservative processes), , and dispersion-diffusion.

6.1.1 Advection

As mentioned in Chapter 4, advection occurs whenever the fluid velocity is different from zero and its intensity depends on the magnitude of $\underline{v}$. This phenomenon, or process, is due to the fact that the dissolved substance is carried by the fluid particles as they move. We observe that in the case of transport by a fluid in a porous medium, two kinds of velocities were defined in Chapter 5: the velocity of fluid particles, $\underline{v}$, and the Darcy velocity, $\underline{U}$. Care must be exercised to distinguish between these two velocities; they are related by

$$\underline{U} = \varepsilon \underline{v} \qquad (6.5)$$

or

$$\underline{v} = \varepsilon^{-1} \underline{U}. \qquad (6.6)$$

6.2 NON-CONSERVATIVE PROCESSES

Mass generation of the solutes may occur by chemical reactions, radioactive decay, and adsorption of the solute by the solid matrix. Adsorption is a well-known phenomenon identified with solute transport in a porous medium that does not occur in a free fluid.

6.2.1 First-order irreversible processes

A convenient notation to express the mass source in the case when the fluid is contained in a porous medium is

$$g_S \equiv \varepsilon \breve{g}_S. \tag{6.7}$$

Here g_S represents the solute mass source per unit time, per unit volume of the physical space occupied by the solid-fluid system, and $\breve{g}_S$ is the mass source per unit time, per unit volume of the fluid. Then, for first-order chemical reactions and radioactive decay, $\breve{g}_S$ is given by Eq. (4.29); that is,

$$g_S \equiv \varepsilon \breve{g}_S = -\lambda \varepsilon c. \tag{6.8}$$

6.2.2 Adsorption

Adsorption is a phenomenon that occurs due to the presence of the solid matrix when the transport of solutes takes place in a porous medium. Actually, this process is a type of chemical reaction in which the solute interacts with the solid phase of the porous medium. A precise and rigorous treatment of adsorption, which takes place in the fluid-solid system, requires approaching it as a system made up of two phases and two components; the phases are the solid matrix and the fluid contained in it. The components of the porous medium are the solute dissolved in the fluid and the solute contained on the solid substance that constitutes the solid matrix. Thus, we have to build a model in which two components are contained by the solid phase while one is contained by the fluid. The following notation will be used for building such a model:

- g_S^f, the solute mass source that goes from the solid phase to the fluid phase;

- g_f^S, the solute mass source that goes from the fluid phase to the solid phase;

- ρ_S, the density of the solid phase (mass of solid over volume of solid);

- $\rho_b \equiv (1 - \varepsilon) \rho_S$, the *bulk density* of the solid phase (mass of solid over total volume of porous medium); and

- ϖ, the *mass fraction* of the solute in the solid phase (mass of solute over mass of solid).

We observe that g_S^f is the only *solute mass source* in the fluid that is due to adsorption; that is, if no other non-conservative processes were present we would have $g_S = g_S^f$ in Eq. (6.3).

The family of extensive properties on which the mathematical model of the adsorption process is based has three members: namely,

(1) the mass of solute dissolved in the fluid, $M_S(t)$; (2) the mass of the substance that constitutes the solid matrix, $M_M(t)$; and (3) the mass of solute adsorbed to the solid matrix, $M_{Ss}(t)$. They are given by Eq. (6.1), together with

$$M_M(t) \equiv \int_{B(t)} \rho_b \, dx = \int_{B(t)} (1 - \varepsilon) \rho_S \, dx \qquad (6.9)$$

and

$$M_{Ss}(t) \equiv \int_{B(t)} \rho_b \varpi \, dx = \int_{B(t)} (1 - \varepsilon) \rho_S \varpi \, dx \qquad (6.10)$$

respectively. The balance equations corresponding to these extensive properties are Eq. (6.2) with $g_s = g_s^f$, together with

$$\frac{dM_M}{dt}(t) = 0 \qquad (6.11)$$

and

$$\frac{dM_{Ss}}{dt}(t) = \int_{B(t)} g_f^S \, dx. \qquad (6.12)$$

Equation (6.11) expresses the conservation of mass of the substance that constitutes the solid matrix. Furthermore, $g_f^S + g_S^f = 0$, due to the requirement of the conservation of mass of the solute contained in the whole system.

Equations (6.11) and (6.12) are, respectively, equivalent to

$$\frac{\partial \rho_b}{\partial t} + \nabla \cdot \left(\rho_b \underline{v}^S\right) = 0 \qquad (6.13)$$

and

$$\frac{\partial \varpi \rho_b}{\partial t} + \nabla \cdot \left(\varpi \rho_b \underline{v}^S\right) = g_f^S. \qquad (6.14)$$

In the presence of Eq. (6.13), Eq. (6.14) can be replaced by

$$\rho_b \left(\frac{\partial \varpi}{\partial t} + \underline{v}^S \cdot \nabla \varpi\right) = g_f^S. \qquad (6.15)$$

Furthermore, $\underline{v}^S = 0$ since the solid matrix is at rest. Hence

$$g_S^f = -g_f^S = -\rho_b \frac{\partial \varpi}{\partial t}. \qquad (6.16)$$

When the mass fraction of the solute in the solid phase is a function of the solute concentration in the fluid, as it is in many applications [10], that is,

$$\varpi \equiv \varpi(c) \qquad (6.17)$$

Eq. (6.16) becomes

$$g_S^f = -\rho_b \frac{d\varpi}{dc} \frac{\partial c}{\partial t}. \qquad (6.18)$$

The *distribution coefficient* is defined to be

$$K_d \equiv \frac{d\varpi}{dc}.$$ (6.19)

In the special case when $\varpi(c)$ is a linear function, the distribution coefficient is independent of the solute concentration. When these observations are incorporated in Eq. (6.3), one obtains

$$\frac{\partial \varepsilon c}{\partial t} + \nabla \cdot (\varepsilon c \underline{v}) = -\rho_b K_d \frac{\partial c}{\partial t} + \nabla \cdot \underline{\tau}_S.$$ (6.20)

In applications, this equation is usually written as

$$R \frac{\partial c}{\partial t} + \varepsilon^{-1} \nabla \cdot (\varepsilon c \underline{v}) = \varepsilon^{-1} \nabla \cdot \underline{\tau}_S - c \frac{\partial \text{Ln} \varepsilon}{\partial t}$$ (6.21)

where Ln is the natural logarithm and R is the *retardation coefficient*, which is defined by

$$R \equiv 1 + \frac{\rho_b}{\varepsilon} \frac{d\varpi}{dc} = 1 + \frac{\rho_b}{\varepsilon} K_d.$$ (6.22)

It is helpful to consider the motivation behind the nomenclature just introduced. When a porous medium is homogeneous and incompressible, Eq. (6.21) is

$$R \frac{\partial c}{\partial t} + \nabla \cdot (c \underline{v}) = \nabla \cdot \underline{\tau}_S$$ (6.23)

which can be written as

$$\frac{\partial c}{\partial t} + \nabla \cdot \left(c R^{-1} \underline{v} \right) = \nabla \cdot \left(R^{-1} \underline{\tau}_S \right).$$ (6.24)

Furthermore, if no adsorption occurs (that is, $K_d = 0$), Eq. (6.24) becomes

$$\frac{\partial c}{\partial t} + \nabla \cdot (c \underline{v}) = \nabla \cdot \underline{\tau}_S.$$ (6.25)

Comparing Eqs. (6.24) and (6.25), it is seen that both are transport models with advection velocities equal to $R^{-1} \underline{v}$ and $\underline{v}$, respectively; that is, *when adsorption occurs the advection process is retarded by a factor equal to R.*

6.3 DISPERSION-DIFFUSION

In the case of solute transport in a porous medium, diffusion processes are also modeled by means of Fick's law; however, for this kind of application it is usually written in the form

$$\underline{\tau}_s(\underline{x}, t) = \varepsilon \underline{\underline{D}} \nabla c.$$ (6.26)

Furthermore, a fundamental difference between diffusion processes that occur in a free fluid and those that take place in a fluid contained in a porous medium is that in the latter two different kinds of phenomena coexist:

1. *Molecular diffusion*, which is due to Brownian motion occurring at the microscopic level; and

2. *Mechanical dispersion*, which is due to the tortuous fluid flow attributable to the randomness of the solid-matrix structure.

Each of these phenomena contributes to the overall dispersion-diffusion process where they have an additive effect. Indeed, the *dispersion tensor* $\underline{\underline{D}}$ is expressed as the sum of the tensor $\underline{\underline{D}}^m$ of *molecular diffusion* plus the tensor $\underline{\underline{D}}^M$ of *mechanical dispersion*; that is,

$$\underline{\underline{D}} = \underline{\underline{D}}^M + \underline{\underline{D}}^m. \tag{6.27}$$

Mechanical dispersion does not occur when the fluid is at rest, but when the fluid is in motion the effect of *mechanical dispersion* is usually more significant than that of *molecular diffusion*.

The *molecular diffusion tensor*, $\underline{\underline{D}}^m$, as given in most text books and used in practical applications, is an isotropic tensor defined by

$$D_{ij}^m = D_d \theta \delta_{ij}. \tag{6.28}$$

Here, θ is a real-number coefficient always less than 1, referred to as the *tortuosity*, which depends on the structure of the solid matrix. It should be observed that when Eq. (6.28) is adopted, one is making the implicit assumption that tortuosity is an isotropic property. However, in more general situations in which the structure of the solid-matrix may be anisotropic and have preferred directions, such an assumption is not valid. However, as noted earlier, such effects are neglected in text books and usual applications.

As for the tensor of mechanical dispersion, $\underline{\underline{D}}^M$, it is anisotropic, having the fluid particle velocity as a preferred direction. As a matter of fact, the following statement characterizes the tensor of mechanical dispersion: All the proper values of the matrix $\underline{\underline{D}}^M$ are proportional to the magnitude, $|\underline{v}|$, of the fluid velocity, while the direction of $\underline{v}$ is an axis of symmetry of $\underline{\underline{D}}^M$. Then, it can be seen that the matrix $\underline{\underline{D}}^M$ necessarily has the following form:

$$\varepsilon D_{ij}^M = D_T |\underline{v}| \delta_{ij} + (D_L - D_T) \frac{v_i v_j}{|\underline{v}|}. \tag{6.29}$$

The direction of the fluid velocity and any direction transversal (i.e., orthogonal) to it are eigenvectors of $\underline{\underline{D}}^M$, with proper values $D_L |\underline{v}|$ and $D_T |\underline{v}|$, respectively. Furthermore, D_L and D_M are called the *longitudinal and transversal coefficients of mechanical dispersivity*, respectively. Generally, $D_L > D_T > 0$. One can also use Darcy's velocity to express $\underline{\underline{D}}^M$; indeed, Eq. (6.29) is equivalent to

$$D_{ij}^M = D_T |\underline{U}| \delta_{ij} + (D_L - D_T) \frac{U_i U_j}{|\underline{U}|}. \tag{6.30}$$

As for a physical interpretation of Eq. (6.29), the following observations are relevant:

1. When the gradient of the solute concentration has the same direction as the velocity, one has

$$\underline{\underline{D}}^M \nabla c = D_L \, |\underline{v}| \, \nabla c. \tag{6.31}$$

That, is the *effective diffusion coefficient* is $D_L \, |\underline{v}|$.

2. When the gradient of the solute concentration has a direction transverse to the fluid velocity, one has

$$\underline{\underline{D}}^M \nabla c = D_T \, |\underline{v}| \, \nabla c. \tag{6.32}$$

That is, the *effective diffusion coefficient* is $D_T \, |\underline{v}|$.

It is also relevant to evaluate the mass flow across a point on the boundary of a body due to mechanical dispersion. We consider two cases:

1. The fluid velocity is perpendicular to the boundary. Then the unit normal vector to the boundary is parallel to the fluid velocity and

$$\underline{\underline{\tau}}_{mec} \cdot \underline{n} = \underline{n} \cdot \varepsilon \underline{\underline{D}}^M \nabla c = \varepsilon D_L \, |\underline{v}| \, \frac{\partial c}{\partial n}. \tag{6.33}$$

2. The fluid velocity is parallel to the boundary. Then, the unit normal vector to the boundary is perpendicular to the fluid velocity and

$$\underline{\underline{\tau}}_{mec} \cdot \underline{n} = \underline{n} \cdot \varepsilon \underline{\underline{D}}^M \nabla c = \varepsilon D_T \, |\underline{v}| \, \frac{\partial c}{\partial n}. \tag{6.34}$$

To finish this analysis, we consider the case when the fluid is at rest. Then

$$D_{ij} = D_{ij}^m = \theta D_d \delta_{ij}. \tag{6.35}$$

As in the case of a free fluid, we have only isotropic molecular diffusion, but it is modified by the tortuosity of the porous medium.

6.4   THE EQUATIONS FOR TRANSPORT OF SOLUTES IN POROUS MEDIA

The basic differential equation, in a general form, that governs the transport of solutes in a porous medium can be obtained by substitution of the diffusive flux of solute, given by Eq. (6.26), into Eq. (6.3):

$$\frac{\partial \varepsilon c}{\partial t} + \nabla \cdot (\varepsilon c \underline{v}) = \nabla \cdot \left(\varepsilon \underline{\underline{D}} \nabla c \right) + g_S. \tag{6.36}$$

This equation can be transformed into

$$\varepsilon c \left(\frac{D \ln c}{Dt} + \frac{D \ln \varepsilon}{Dt} + \nabla \cdot \underline{v} \right) = \nabla \cdot \left(\varepsilon \underline{\underline{D}} \nabla c \right) + g_S. \tag{6.37}$$

This, in turn, can be transformed into

$$\frac{Dc}{Dt} + c \left(\frac{D \ln \varepsilon}{Dt} + \nabla \cdot \underline{v} \right) = \nabla \cdot \left(\underline{\underline{D}} \nabla c \right) + (\nabla \ln \varepsilon) \cdot \underline{\underline{D}} \nabla c + \varepsilon^{-1} g_S. \tag{6.38}$$

This is a very general form of such a governing equation, and applications of it to some particular cases are derived next.

When the fluid is incompressible, in view of Eq. (5.2) on page 87, one has

$$\frac{\partial \varepsilon}{\partial t} + \nabla \cdot (\varepsilon \underline{v}) = 0 \tag{6.39}$$

or, equivalently,

$$\frac{D \ln \varepsilon}{Dt} + \nabla \cdot \underline{v} = 0. \tag{6.40}$$

Hence,

$$\frac{Dc}{Dt} = \nabla \cdot \left(\underline{\underline{D}} \nabla c \right) + (\nabla \ln \varepsilon) \cdot \underline{\underline{D}} \nabla c + \varepsilon^{-1} g_S. \tag{6.41}$$

If the solid matrix is homogeneous (actually, if $\nabla \ln \varepsilon$ can be neglected),

$$\frac{Dc}{Dt} = \nabla \cdot \left(\underline{\underline{D}} \nabla c \right) + \varepsilon^{-1} g_S. \tag{6.42}$$

Or, more explicitly,

$$\frac{\partial c}{\partial t} + \underline{v} \cdot \nabla c = \nabla \cdot \left(\underline{\underline{D}} \nabla c \right) + \varepsilon^{-1} g_S. \tag{6.43}$$

When the fluid is at rest, this equation reduces to

$$\frac{\partial c}{\partial t} = \nabla \cdot (\theta D \nabla c) + \varepsilon^{-1} g_S. \tag{6.44}$$

For non-diffusive transport, Eq. (6.43) reduces to

$$\frac{\partial c}{\partial t} + \underline{v} \cdot \nabla c = \varepsilon^{-1} g_S. \tag{6.45}$$

All these governing equations, Eqs. (6.43) to (6.45), are used extensively in the modeling of groundwater contamination. They show explicitly that, as in transport by a free fluid, in transport of solutes by a fluid in a porous medium the advection velocity is also the particle velocity of the fluid. Furthermore, Eq. (6.43) is very similar to Eq. (4.7), which governs solute transport by an incompressible free fluid; in particular, it shares with the latter equation the very important property of being a parabolic equation. However, in Eq. (6.43) the *diffusion-dispersion matrix* is generally anisotropic, while in Eq. (4.7) that matrix is necessarily isotropic. As for Eq. (6.45), which governs *non-diffusive transport*, it is a first order equation.

The equations governing steady states that stem from Eqs. (6.43) and (6.44) are

$$\underline{v} \cdot \nabla c = \nabla \cdot \left(\underline{\underline{D}} \nabla c \right) + \varepsilon^{-1} g_S \tag{6.46}$$

$$\nabla \cdot (\theta D \nabla c) = -\varepsilon^{-1} g_S \tag{6.47}$$

and

$$\underline{v} \cdot \nabla c = \varepsilon^{-1} g_S. \tag{6.48}$$

6.5 WELL-POSED PROBLEMS

As has been said repeatedly, the kind of problems that are well posed for each partial differential equation is determined by its type. As for the governing equations of solute transport by a fluid in a porous medium, their corresponding types are: parabolic for Eqs. (6.43) and (6.44); elliptic for Eqs. (6.46) and (6.47); and first order for Eqs. (6.45) and (6.48). The corresponding well-posed problems are essentially the same as those that occur in solute transport by a free fluid and that were discussed in Sections 4.5 and 4.8. To avoid excessive repetition, the reader is referred to those sections.

6.6 SUMMARY

In this chapter we considered the transport of a dissolved solute in a porous medium. First the general balance equation was developed. Next, the processes of advection, hydrodynamic dispersion, and diffusion were presented. Transport was divided into conservative and non-conservative. In the case of non-conservative transport, first-order irreversible reactions and the phenomenon of adsorption, which is unique to porous medium flow, were considered. To close the system of equations, constitutive relations were introduced to describe the first-order reactions, adsorption, and dispersive and diffusive transport. The chapter concluded with a discussion of well-posed problems.

EXERCISES

6.1   When the fluid contained in a porous medium is incompressible, Eq. (1.43) holds:

$$\frac{\partial \varepsilon}{\partial t} + \nabla \cdot (\varepsilon \underline{v}) = 0. \tag{6.49}$$

Write Eq. (6.3), which governs the transport of a substance dissolved in a fluid contained in a porous medium, for the conservative and non-diffusive case ($g_S = 0$,

$\underline{\tau}_S = 0$) and show that when the fluid is incompressible the fluid particles conserve their solute concentration during their motion; i.e., $Dc/Dt = 0$.

6.2   Establish the following result, which is similar to that of Exercise 6.1. Assume that the fluid mass is conserved; i.e.,

$$\frac{\partial \varepsilon \rho}{\partial t} + \nabla \cdot (\varepsilon \rho \underline{v}) = 0. \tag{6.50}$$

Combining this equation with Eq. (6.3), with $g_S = 0$ and $\underline{\tau}_S = 0$, show that the *mass fraction of the solute in the fluid phase*, $\omega \equiv c/\rho'$, is conserved by the fluid particles during their motion; i.e., $D\omega/Dt = 0$.

6.3   Write Eq. (6.3), with $g_S = -\lambda \varepsilon c$ and $\underline{\tau}_S = 0$, to obtain

$$\frac{\partial \varepsilon c}{\partial t} + \nabla \cdot (\varepsilon c \underline{v}) + \lambda \varepsilon c = 0. \tag{6.51}$$

Show that in this case the fluid particles conserve $e^{\lambda \varepsilon t} \omega$; i.e., on each fluid particle the *mass fraction of the solute* decays according to the factor $e^{-\lambda \varepsilon t}$.

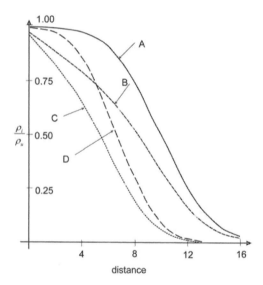

Figure 6.1   Concentration profiles involving retardation and chemical reactions.

6.4   Assume that all the curves shown in Fig. 6.1 are observed at the same time: namely 20 sec. Each has a combination of the following: $R = 1$, $R = 1.5$, $\lambda = 0.05$, $\lambda = 0.0$, where R is the retardation and λ the reaction rate. Assume that the groundwater gradient is -0.1 cm/cm, that the hydraulic conductivity is 1.0 cm/sec, and that the porosity ε is 0.2. Hint: $v = -\frac{K}{\varepsilon} \frac{dh}{dx}$
   a) Which curve has $R = 1$, $\lambda = 0$, and why?

b) Which curve corresponds to $R = 1.5$, $\lambda = 0$, and why?
c) Which curve corresponds to $R = 1.0$, $\lambda = 0.05$, and why?

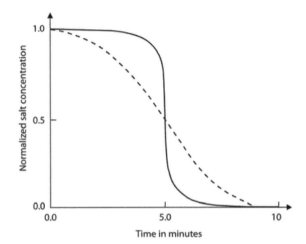

Figure 6.2 The vertical axis is the normalized concentration, i.e., the concentration divided by the highest concentration value. The horizontal axis is the time at which observations of concentration were observed at the end of the column.

6.5 In Fig. 6.2 there are two concentration versus time curves, one for each of two experiments conducted on the same column but packed with different materials with the same permeability and porosity. The curves were observed by examining the concentration at the ends of the column. The length of the column is 20 cm.

a) Which of the two curves shows the greater dispersion, and why?
b) Assuming no retardation, what was the pore velocity of the salt solution? Show how you calculated it.

6.6 It is observed in the field that the dispersivity increases with distance traveled by a solute. Provide a plausible explanation for this observation.

REFERENCES

1. Abriola, L.M., *Multiphase Migration of Organic Compounds in a Porous Medium: A Mathematical Model*, Lecture Notes in Engineering Vol.8, Eds. C.A. Brebbia and S.A. Orszag, Springer-Verlag, Berlin, 1988.

2. Anderson, M.P. and W.W. Woessner, *Applied Groundwater Modeling: Simulation of Flow and Advective Transport*, Academic Press, San Diego, CA, 1991.

3. Bear, J., *Dynamics of Fluids in Porous Media*, Elsevier, New York, 1972.

4. Bird, B.R., W.E. Stewart, and E.N. Lightfoot, *Transport Phenomena*, Wiley, New York, 1960.

5. Green, A.E., W. Green-Zerna, and W. Zerna, *Theoretical Elasticity*, reprint edition, Dover, New York, 1992.

6. Huyakorn, P.S. and G.F. Pinder, *Computational Methods in Subsurface Flow*, Academic Press, New York, 1983.

7. Lee, T.C., *Applied Mathematics in Hydrogeology*, Lewis Publishers, Boca Raton, FL, 1999.

8. National Research Council, *Ground Water Models: Scientific and Regulatory Applications*, National Academy Press, Washington, DC, 1990

9. Spitz, K. and J. Moreno, *A Practical Guide to Groundwater and Solute Transport Modeling*, Wiley, New York, 1996.

10. Zheng, C.H. and G.D. Bennett, *Applied Contaminant Transport Modeling*, Van Nostrand Reinhold, Wiley, 1995.

CHAPTER 7

MULTIPHASE SYSTEMS

In Chapters 5 and 6 we introduced the concept of fluid flow through a porous medium within the context of a system consisting of two phases: the solid phase or grains and a fluid phase (water). We now turn our attention to a more complex porous medium system that has multiple fluid phases. Such models are used primarily in the study of petroleum and geothermal reservoir engineering, soil physics, and contaminant hydrology. This chapter is devoted to the study of models that are mainly relevant in contaminant hydrology and soil physics, and in the next chapter our attention will be centered on petroleum reservoir models.

7.1 BASIC MODEL FOR THE FLOW OF MULTIPLE-SPECIES TRANSPORT IN A MULTIPLE-FLUID- PHASE POROUS MEDIUM

The axiomatic procedure for developing the basic mathematical models of multiphase systems was presented in Sections 3.1 and 3.2. The approach that we will take to develop the equations describing the flow of multiple fluid phases through a porous medium begins with a description of the transport of individual dissolved species and then combines these equations to yield the fluid-flow equations. In Chapter 4

Mathematical Modeling in Science and Engineering: An Axiomatic Approach.
By Ismael Herrera and George F. Pinder Copyright © 2012 John Wiley & Sons, Inc.

we introduced the concept of species transport in a free fluid, and in Chapter 6 we expanded this discussion to include transport by a fluid that flows through a porous medium.

The concepts of phase and component were introduced in Chapter 3. Here, the term *species* will be used as synonymous with component; i.e., a species is a substance that is dissolved in a phase. The phases to be considered in the present chapter will be solid, liquid, or gas, which at the microscopic level will be separated from another solid, liquid, or gas phase by a boundary or interface. Each phase will be identified with the notation α. In some models of such systems, contrary to what we did in Chapters 5 and 6, the solid phase will be accounted for separately; then, using the definitions above, a partially saturated soil would contain three phases: a water phase ($\alpha = w$), a gas phase (air; $\alpha = g$), and a solid phase (the grains; $\alpha = s$). Such a system is presented in Fig. 7.1.

The phases of such a system are separated from each other and therefore they occupy different parts of the physical space; thus, each phase has its own volume V^{α}. The volume *fraction of phase* α is defined to be $\varepsilon^{\alpha} \equiv V^{\alpha}/V$, where V^{α} is the volume occupied by phase α and V is the total volume of the system.

Each fluid phase contains dissolved species and we will denote species i in phase α by adorning the appropriate state variable with the notation αi. For example, the mass fraction of species i in phase α will read $\omega^{\alpha i}$. The mass fraction is related to the concentration, defined as mass of i per unit volume of phase α, that is, $\rho^{i\alpha}$, through the relationship $\rho^{i\alpha} = \rho^{\alpha}\omega^{i\alpha}$, where ρ^{α} is the density of phase α.

We begin with the species transport equation presented in Chapter 6, but now written for a specific phase of a multiphase system.

7.2 MODELING THE TRANSPORT OF SPECIES I IN PHASE α

To apply the axiomatic procedure for establishing the basic models of multiphase systems of (Section 3.2 [Eqs. (3.1) and (3.2) on page 46], we need to identify a suitable family of extensive properties. For all these applications, the mass of each species contained in each phase, for each phase α and each component i, is a member of such a family. Let $M^{i\alpha}$ be the mass of species i contained in phase α, where $i = 1, ..., N_c$ (N_c is the total number of species) and $\alpha = 1, ..., N_p$ (N_p is the total number of phases); then the family of extensive properties characterizing the model is constituted by

$$\left\{ M^{i\alpha}, \quad i = 1, ..., N_c \text{ and } \alpha = 1, ..., N_p \right\}. \tag{7.1}$$

The total number of this family is $N_c \times N_p$. However, this family can be reduced. If species i is not soluble in phase α, we know in advance that $M^{i\alpha} = 0$ and it can be eliminated. Therefore, it is better to develop the mathematical model starting from the reduced family $\left\{ M^{i\alpha} \right\}$, with $i = 1, ..., N_{\alpha C}$ and $\alpha = 1, ..., N_p$. Here, $N_{\alpha C}$ is the number of species whose solubility in phase α is non-vanishing. The cardinality

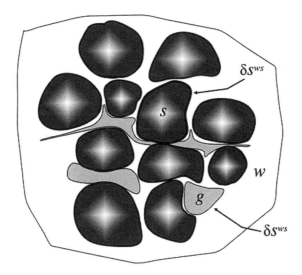

Figure 7.1   Idealized representation of a porous medium containing two fluid phases. In this figure s is solid, g is gas, w is water, δS^{ws} is the interface between the water and solid, and δS^{gw} is the interface between the gas and the water.

(that is, the total number of members) of such a family of extensive properties is

$$\sum_{\alpha=1}^{N_p} N_{\alpha C}.$$
(7.2)

The differential equations governing the transport of multiple species in multiple phases can now be derived from Eqs. (3.1). To this end we express the family of extensive properties just introduced as integrals over the bodies of the multiphase system:

$$M^{i\alpha}(t) = \int_{B(t)} \varepsilon^{\alpha}(\underline{x}, t) \, \rho^{i\alpha}(\underline{x}, t) \, dx.$$
(7.3)

Then the corresponding intensive properties characterizing the model of such a system are the integrands in Eq. (7.3). This yields the following family of intensive properties:

$$\left\{ \varepsilon^{\alpha} \rho^{i\alpha}, \ i = 1, ..., N_{\alpha C} \text{ and } \alpha = 1, ..., N_p \right\}.$$
(7.4)

Furthermore, according to the global balance equation, Eq. (1.25) (on page 10),

$$\frac{dM^{i\alpha}}{dt}(t) = \int_{B(t)} g^{i\alpha}(\underline{x}, t) \, d\underline{x} + \int_{\partial B(t)} \underline{\tau}^{i\alpha}(\underline{x}, t) \cdot \underline{n}(\underline{x}, t) \, d\underline{x}$$
(7.5)

where $g^{i\alpha}(\underline{x}, t)$ are the *external supplies of mass of species i to phase α*, while $\underline{\tau}^{i\alpha}(\underline{x}, t)$ is the *flux of mass of species i to phase α*. The term $g^{i\alpha}$ will be decomposed into three parts: namely, a part $\varepsilon^\alpha r^{i\alpha}$, due to *chemical reactions* occurring in phase α that generate mass of species i; a part $g_I^{i\alpha}$, due to interchange of mass between the different phases; and a part $g_E^{i\alpha}$, coming from the exterior of the transport system. It is important to notice that the way in which we have written these terms implies that $r^{i\alpha}$ *is the mass produced by chemical reactions occurring in phase* α, per unit volume of that phase, per unit time. The term $g_E^{i\alpha}$ is usually due to injection, or extraction, of the fluids through wells. Here, the sign convention is that $g_E^{i\alpha}$ is positive when there is injection, negative when there is extraction, and zero otherwise. Let $e_\beta^{i\alpha}$ be the mass (per unit total volume of the system) of species i that goes from phase α to phase β across the interface that separates these two phases. Then

$$g_I^{i\alpha} = -\sum_{\beta \neq \alpha} e_\beta^{i\alpha} \tag{7.6}$$

and

$$g^{i\alpha} = \varepsilon^\alpha r^{i\alpha} - \sum_{\beta \neq \alpha} e_\beta^{i\alpha} + g_E^{i\alpha}. \tag{7.7}$$

As for the flux of mass of species i to phase α, that is, the vector field $\underline{\tau}^{i\alpha}$ in Eq. (7.5), it is due to diffusion-dispersion processes of the type described in Chapter 6, performed by the species that is dissolved in phase. It is usually written as

$$\underline{\tau}^{i\alpha} = -\varepsilon^\alpha \underline{j}^{i\alpha}. \tag{7.8}$$

The vector $\underline{j}^{i\alpha}$ is referred to as the dispersion vector for species i in phase α.

Application of Eq. (3.1), using Eqs. (7.7) and (7.8), yields

$$\frac{\partial \left(\varepsilon^\alpha \rho^{ia}\right)}{\partial t} + \nabla \cdot \left(\varepsilon^\alpha \rho^{ia} \underline{v}^\alpha\right) + \nabla \cdot \left(\varepsilon^\alpha \underline{j}^{i\alpha}\right) - \varepsilon^\alpha r^{i\alpha} + \sum_{\alpha \neq \beta} e_{\alpha\beta}^{i\alpha} = 0;$$
$$i = 1, ..., N_{\alpha C}, \quad \alpha = 1, ..., N_p. \tag{7.9}$$

For simplicity, when writing this equation we have taken $g_E^{i\alpha} = 0$; that is, we assume that there is no injection or pumping of fluids through wells, and we shall continue to do this in what follows.

It is interesting to write Eq. (7.9) in terms of the mass fraction and then sum over all the species in the phase to give

$$\sum_{i=1}^{N_{\alpha C}} \left[\frac{\partial \left(\varepsilon^\alpha \rho^a \omega^{i\alpha}\right)}{\partial t} + \nabla \cdot \left(\varepsilon^\alpha \rho^a \omega^{i\alpha} \underline{v}^\alpha\right)\right]$$
$$+ \sum_{i=1}^{N_{\alpha C}} \left[\nabla \cdot \left(\varepsilon^\alpha \underline{j}^{i\alpha}\right) - \varepsilon^\alpha r^{i\alpha} + \sum_{\alpha \neq \beta} e_{\alpha\beta}^{i\alpha}\right]$$
$$= 0; \quad \alpha = 1, ..., N_p. \tag{7.10}$$

Moving the summation through the differential operators and noting that $\varepsilon^\alpha \rho^\alpha$ and $\underline{v}^\alpha$ are independent of i, one obtains

$$\frac{\partial \left(\varepsilon^\alpha \rho^a \sum_{i=1}^{N_\alpha C} \omega^{i\alpha} \right)}{\partial t} + \nabla \cdot \left(\varepsilon^\alpha \rho^a \underline{v}^\alpha \sum_{i=1}^{N_\alpha C} \omega^{i\alpha} \right) +$$

$$\nabla \cdot \left(\varepsilon^\alpha \sum_{i=1}^{N_\alpha C} \underline{j}^{i\alpha} \right) - \varepsilon^\alpha \sum_{i=1}^{N_\alpha C} r^{i\alpha} + \sum_{i=1}^{N_\alpha C} \sum_{\alpha \neq \beta} e_{\alpha\beta}^{i\alpha}$$

$$= 0; \quad \alpha = 1, ..., N_p. \tag{7.11}$$

By definition the mass fractions sum to unity, and conservation of mass implies that

$$\sum_{i=1}^{N_\alpha C} r^{i\alpha} = 0. \tag{7.12}$$

Observe that Eq. (7.12) is fulfilled only when reactions occurring within the phase are taken into account. Thus, radioactive decay or first order reactions, for example, must not be incorporated in $r^{i\alpha}$. Furthermore, define

$$\underline{j}^\alpha \equiv \sum_{i=1}^{N_\alpha C} \underline{j}^{i\alpha} \tag{7.13}$$

and

$$e_\beta^\alpha \equiv \sum_{i=1}^{N_\alpha C} e_\beta^{i\alpha}. \tag{7.14}$$

Noting that $\underline{j}^\alpha = 0$, we obtain from Eq. (7.11),

$$\frac{\partial \left(\varepsilon^\alpha \rho^a \right)}{\partial t} + \nabla \cdot \left(\varepsilon^\alpha \rho^a \underline{v}^\alpha \right) + \sum_{\alpha \neq \beta} e_\beta^\alpha = 0, \quad \alpha = 1, ..., N_p. \tag{7.15}$$

This is the equation for movement of the α phase. The last term in this equation describes the mass transfer from phase α to all other phases with which phase α makes contact.

7.3 THE SATURATED FLOW CASE

In this section we discuss the flow of a fluid, such as water, flowing in a porous medium such as a soil. We already dealt with this subject in Chapter 5. However, there it was assumed that there was no interaction between the fluid and the solid. On the contrary, here we analyze the case when the fluid is adsorbed on the solid. Models of this kind have significant applications in soil science and hydrology. As will be seen, Eq. (5.47) will be used once again, but exchange between the fluid and the solid gives rise to a source term. For greater clarity we prefer to develop this model *ab initio*.

Let us begin by writing Eq. (7.15) for a water phase w as

$$\frac{\partial\left(\varepsilon^w\rho^w\right)}{\partial t} + \boldsymbol{\nabla}\cdot\left(\varepsilon^w\rho^w\underline{v}^w\right) - e^w_{ws} = 0 \tag{7.16}$$

and

$$\frac{\partial\left(\varepsilon^s\rho^s\right)}{\partial t} + \boldsymbol{\nabla}\cdot\left(\varepsilon^s\rho^s\underline{v}^s\right) - e^s_{sw} = 0. \tag{7.17}$$

We now define a Darcy velocity $\underline{q}$ as

$$\underline{q} = \varepsilon\underline{v}^w - \varepsilon\underline{v}^s$$

where, as earlier, ε is the porosity. Substitution of this definition into Eq. (7.16) yields

$$\frac{\partial\left(\varepsilon^w\rho^w\right)}{\partial t} + \boldsymbol{\nabla}\cdot\left(\rho^w\left[\underline{q} + \varepsilon^w\underline{v}^s\right]\right) - e^w_{ws} = 0. \tag{7.18}$$

We now multiply Eq. 7.17 by $\frac{\varepsilon^w\rho^w}{\varepsilon^s\rho^s}$ so as to eliminate the grain velocity term and obtain

$$\frac{\partial\left(\varepsilon^w\rho^w\right)}{\partial t} + \boldsymbol{\nabla}\cdot\left(\rho^w\underline{q}\right) + \underline{v}^s\cdot\boldsymbol{\nabla}\left(\varepsilon^w\rho^w\right)$$
$$- e^w_{ws} - \frac{\varepsilon^w\rho^w}{\varepsilon^s\rho^s}\left[\frac{\partial\left(\varepsilon^s\rho^s\right)}{\partial t} + \underline{v}^s\cdot\boldsymbol{\nabla}\left(\varepsilon^s\rho^s\right) - e^s_{sw}\right] = 0. \tag{7.19}$$

Now regroup the terms in this equation to give

$$\frac{D\left(\varepsilon^w\rho^w\right)}{Dt} - \frac{\varepsilon^w\rho^w}{\varepsilon^s\rho^s}\frac{D\left(\varepsilon^s\rho^s\right)}{Dt} + \boldsymbol{\nabla}\cdot\left(\rho^w\underline{q}\right) - e^w_{ws} + \frac{\varepsilon^w\rho^w}{\varepsilon^s\rho^s}e^s_{sw} = 0 \tag{7.20}$$

where the material derivative $\frac{D(\cdot)}{Dt}$ is taken with respect to the velocity of the solid particles.

Following the development of Pinder and Gray[17], we expand the first term in this equation to yield

$$\frac{D\left(\varepsilon^w\rho^w\right)}{Dt} = \rho^w\frac{D\varepsilon^w}{Dt} + \varepsilon^w\frac{D\rho^w}{Dt}. \tag{7.21}$$

The second term in Eq. (7.21) can be written in the equivalent form

$$\rho^w\frac{D\varepsilon^w}{Dt} = -\rho^w\frac{D\left(1 - \varepsilon^w\right)}{Dt}$$
$$= -\frac{\rho^w}{\rho^s}\left[\rho^s\frac{D\left(\varepsilon^s\right)}{Dt}\right]$$
$$= -\frac{\rho^w}{\rho^s}\left[\frac{D\varepsilon^s\rho^s}{Dt} - \varepsilon^s\frac{D\rho^s}{Dt}\right]. \tag{7.22}$$

Now substitute Eq. (7.22) into Eq. (7.20) to give

$$\varepsilon^w \frac{D\rho^w}{Dt} + \varepsilon^s \frac{\rho^w}{\rho^s} \frac{D\rho^s}{Dt} - \frac{\rho^w}{\rho^s \varepsilon^s} \frac{D\left(\varepsilon^s \rho^s\right)}{Dt} + \mathbf{\nabla} \cdot \left(\rho^w \underline{q}\right) - e_{ws}^w + \frac{\varepsilon^w \rho^w}{\varepsilon^s \rho^s} e_{sw}^s = 0. \quad (7.23)$$

Next we use the chain rule to obtain

$$
\begin{aligned}
\varepsilon^w \frac{D\rho^w}{Dt} &= \varepsilon^w \frac{\partial \rho^w}{\partial p^w} \frac{Dp^w}{Dt} \\
&= \varepsilon^w \rho^w \beta^w \frac{Dp^w}{Dt}
\end{aligned}
\quad (7.24)
$$

where β^w is the *fluid compressibility*. Similarly, we can address the compressibility of the solid grains and obtain

$$
\begin{aligned}
\varepsilon^s \frac{\rho^w}{\rho^s} \frac{D\rho^s}{Dt} &= \varepsilon^s \frac{\rho^w}{\rho^s} \frac{\partial \rho^s}{\partial p^w} \frac{Dp^w}{Dt} \\
&= \varepsilon^s \rho^w \beta^s \frac{Dp^w}{Dt}.
\end{aligned}
\quad (7.25)
$$

Finally, we consider the third term in Eq. (7.23):

$$
\begin{aligned}
\frac{\rho^w}{\rho^s \varepsilon^s} \frac{D\left(\varepsilon^s \rho^s\right)}{Dt} &= \frac{\rho^w}{\rho^s \varepsilon^s} \frac{\partial \left(\varepsilon^s \rho^s\right)}{\partial p^w} \frac{Dp^w}{Dt} \\
&= \rho^w \alpha^b \frac{Dp^w}{Dt}
\end{aligned}
\quad (7.26)
$$

where

$$\alpha^b \equiv \frac{-1}{\rho^s \varepsilon^s} \frac{\partial \left(\varepsilon^s \rho^s\right)}{\partial p^w} \quad (7.27)$$

is called the *bulk compressibility*.

Let us now substitute Eqs. (7.24) -(7.26) into Eq. (7.23); we obtain

$$\rho^w \left(\varepsilon^w \beta^w + \varepsilon^s \beta^s + \alpha^b\right) \frac{Dp^w}{Dt} + \mathbf{\nabla} \cdot \left(\rho^w \underline{q}\right) - e_{ws}^w + \frac{\varepsilon^w \rho^w}{\varepsilon^s \rho^s} e_{sw}^s = 0. \quad (7.28)$$

If we now substitute Darcy's law, that is,

$$\underline{q} = -\frac{\underline{\underline{k}}}{\mu} \cdot \left(\mathbf{\nabla} p^w - \rho^w \underline{g}\right) \quad (7.29)$$

and define the *storage coefficient* as

$$S_s = \rho^w g \left(\varepsilon^w \beta^w + \varepsilon^s \beta^s + \alpha^b\right) \quad (7.30)$$

we obtain

$$\frac{S_s}{g} \frac{Dp^w}{Dt} - \mathbf{\nabla} \left[\frac{\rho^w \underline{\underline{k}}}{\mu} \left(\mathbf{\nabla} p^w - \rho^w \underline{g}\right)\right] - e_{ws}^w + \frac{\varepsilon^w \rho^w}{\varepsilon^s \rho^s} e_{sw}^s = 0 \quad (7.31)$$

which is the equation that can be used to describe saturated groundwater flow when there is a water-mass exchange between the water phase and the solid phase. If it is assumed that the matrix deforms so slowly that the substantial derivative can be replaced by a partial derivative with respect to time, one obtains

$$\frac{S_s}{g}\frac{\partial p^w}{\partial t} - \boldsymbol{\nabla}\cdot\left[\frac{\rho^w\underline{\underline{k}}}{\mu}\cdot\left(\boldsymbol{\nabla}p^w - \rho^w\underline{g}\right)\right] - e^w_{ws} + \frac{\varepsilon^w\rho^w}{\varepsilon^s\rho^s}e^s_{sw} = 0. \tag{7.32}$$

This equation can be simplified for systems where the density depends only on pressure and the density gradient is small by recalling the definition of *hydraulic head,* Eq. (5.32) on page 92, which is

$$h^w = \int_{P_{ref}}^{p^w}\frac{dp^{w'}}{\rho\left(p^{w'}\right)g} - \frac{\underline{g}\cdot\underline{k}}{g}\left(z - z_{ref}\right). \tag{7.33}$$

Using *Leibnitz's rule* for differentiation of an integral, we get

$$\frac{\partial h^w}{\partial t} = \frac{1}{\rho^w g}\frac{\partial p^w}{\partial t} \tag{7.34}$$

and

$$\boldsymbol{\nabla}h^w = \frac{1}{\rho^w g}\boldsymbol{\nabla}p^w - \frac{\underline{g}}{g}. \tag{7.35}$$

Substitution of these two derivatives into Eq. (7.32) yields

$$S_s\frac{\partial h^w}{\partial t} - \boldsymbol{\nabla}\cdot\left[\left(\rho^w g\frac{\underline{\underline{k}}}{\mu}\right)\cdot\boldsymbol{\nabla}h^w\right] - \frac{1}{\rho^w}\left(e^w_{ws} + \frac{\varepsilon^w\rho^w}{\varepsilon^s\rho^s}e^s_{sw}\right) = 0 \tag{7.36}$$

or

$$S_s\frac{\partial h^w}{\partial t} - \boldsymbol{\nabla}\cdot\left[\underline{\underline{K}}\cdot\boldsymbol{\nabla}h^w\right] - \frac{1}{\rho^w}\left(e^w_{ws} + \frac{\varepsilon^w\rho^w}{\varepsilon^s\rho^s}e^s_{sw}\right) = 0 \tag{7.37}$$

where $\underline{\underline{K}} \equiv \rho^w g\frac{\underline{\underline{k}}}{\mu}$ is the *hydraulic conductivity.* Recognizing that the flux across phase boundaries must balance, that is,

$$e^w_{ws} + e^s_{sw} = 0 \tag{7.38}$$

we obtain

$$S_s\frac{\partial h^w}{\partial t} - \boldsymbol{\nabla}\cdot\left[\underline{\underline{K}}\cdot\boldsymbol{\nabla}h^w\right] - \frac{1}{\rho^w}e^w_{ws}\left(1 - \frac{\varepsilon^w\rho^w}{\varepsilon^s\rho^s}\right) = 0. \tag{7.39}$$

This formulation has the advantage of using the hydraulic head which is easily measured in the field as the state variable, which, in turn, facilitates the specification of boundary and initial conditions.

In the next section we will consider the multiphase case.

7.4 THE AIR-WATER SYSTEM

We will first describe air-water flow in a soil since it is of practical importance, especially in agronomy, and is a relatively straight forward example. In Section 3.6 we introduced the concept of saturation. If we define the porosity of the porous medium as ε, then $\varepsilon^\alpha = V_\alpha/V_v \times V_v/V = S^\alpha \times \varepsilon$, where S^α is the *saturation of phase phase* α. Following the same general development as that provided above [17], one can obtain for the two-fluid system the equation

$$
S^\alpha \varepsilon \frac{D\rho^\alpha}{Dt} + \varepsilon \rho^\alpha \frac{DS^\alpha}{Dt} + S^\alpha \rho^\alpha \frac{D\varepsilon}{Dt}
$$
$$
+ S^\alpha \varepsilon \rho^\alpha \boldsymbol{\nabla} \cdot \underline{v}^s + \boldsymbol{\nabla} \cdot \left(\rho^\alpha \underline{q}^\alpha \right) - e^\alpha_{\alpha s} - e^\alpha_{wa}
$$
$$
= \quad 0; \quad \alpha = w, a \tag{7.40}
$$

which, when combined with the solid-phase equation for the term $\boldsymbol{\nabla} \cdot \underline{v}^s$, gives

$$
-\frac{S^\alpha \rho^\alpha}{(1-\varepsilon) \rho^s} \frac{D\left[(1-\varepsilon)\rho^s\right]}{Dt} + \frac{S^\alpha \rho^\alpha (1-\varepsilon)}{\rho^s} \frac{D\rho^s}{Dt} + \varepsilon \rho^\alpha \frac{DS^\alpha}{Dt} + S^\alpha \varepsilon \frac{D\rho^\alpha}{Dt}
$$
$$
+ \boldsymbol{\nabla} \cdot \left(\rho^\alpha \underline{q}^\alpha \right) - e^\alpha_{\alpha s} - e^\alpha_{wa} + \frac{\rho^\alpha \varepsilon S^\alpha}{\rho^s (1-\varepsilon)} \left(e^s_{ws} + e^s_{as} \right)
$$
$$
= \quad 0; \quad \alpha = w, a. \tag{7.41}
$$

Experimental relationships exist that relate the saturation of phase α to the pressure in the phase α. A typical example is shown in Fig. 7.2. The ordinate is the *capillary pressure head,* which is defined as $p_c/\rho_w g$, where $p_c = p_w - p_a$ in the air-water system. Note that the capillary pressure is a negative number since atmospheric pressure is traditionally defined to be zero.

Consider now the various curves shown in Fig. 7.2. Let us begin at complete saturation ($S^w = 1$) and a capillary pressure of zero ($p_c = 0$). As the pressure is reduced, that is, as p_c increases, there initially is no change in saturation. At a capillary pressure head of about 5 cm, a decrease in saturation begins as air enters the porous medium. The pressure at which this takes place is called the *bubbling pressure.* As the pressure is further reduced, the saturation decreases until it forms an asymptote at a saturation of approximately 0.3; this saturation is called the *irreducible saturation.* It is at this point that the continuity of the water is compromised and water pressure can no longer be propagated through the porous medium. This curve is the *primary drainage curve,* denoted in Fig. 7.2 as PDC.

If the process is now reversed, that is, we begin to increase the water pressure (decrease p_c) the saturation of water increases. However, it does not return to full saturation because of trapped air in the porous medium. The saturation at which this takes place is called the *residual air saturation* (about 0.82 in this example). This curve is called the *main imbibition curve* and denoted on Fig. 7.2 as MIC.

Now we decrease the water pressure again. The result is a decrease in saturation until the same asymptote is reached at the irreducible water saturation. This curve is called the *main drainage curve* and denoted in Fig. 7.2 as MDC.

Note that at any saturation, but for irreducible and residual values, two pressures may exist. The different curves reflect the drainage or imbition history of the sample. This phenomenon is called *hysteresis*.

One can also begin with a totally dry soil and begin to introduce water. The water pressure increases as water enters the soil. Saturation also increases until residual air saturation is realized. The resulting curve is called the *primary imbibition curve* and denoted on Fig. 7.2 as PIC.

The curves denoted by MDC* and MIC* indicate the evolution of the system if initial saturations are between zero and irreducible water and residual saturation and unity.

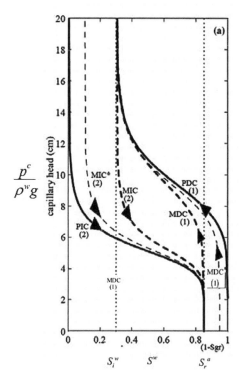

Figure 7.2   Definition plot of the hysteretic relationship between saturation and capillary pressure. Curve position and shape are governed by the mobility of the phases, the initial phase saturations when the process begins, and whether the wetting phase is draining (D) or imbibing (I). The curve-type numbering is such that odd numbers relate to drainage and even numbers to imbibition. Primary (P) and main (M) curves are designated as \1" for drainage processes begun with $1 - s_r^n \leq s^w \leq 1$ and \2" for imbibition processes begun with $0 \leq s^w \leq s_i^w$. Scanning (S) curves indicate behavior after a drainage process is reversed (curves 4 and 6) or an imbibition process is reversed (curves 3 and 5) such that the initial saturation of the new process is $s_i^w < s^w < 1 - s_r^n$.

A second constitutive relationship we require relates permeability to degree of saturation. The coexistence of water and air decreases the permeability to both. Thus the form of Darcy's law requires modification to recognize this physical reality. The result is the following:

$$\underline{q}^{\alpha} = -\frac{k_{r\alpha}\underline{\underline{k}}}{\mu}\left(\nabla p_{\alpha}+\rho\hat{g}\nabla z\right) \tag{7.42}$$

where $k_{r\alpha}$ is the *relative permeability* with a range of zero to 1. The more phase α present in the system, the higher is the value of $k_{r\alpha}$. The relationships between relative permeability and saturation in the case of primary drainage and main drainage and imbibition are provided in Fig. 7.3.

Consider first the case of primary drainage (the left-hand panel in Fig. 7.3). The relative permeability of water is unity at full water saturation. As the water saturation approaches the irreducible saturation, the relative permeability to water approaches zero. At irreducible saturation the water phase is no longer continuous, pressure cannot be propagated, and no flow is possible.

On the other hand, the relative permeability of air is zero at full water saturation and increases as the water saturation decreases. It reaches a maximum at the irreducible saturation of water. It is not unity because of the existence of the irreducible saturation of water.

If the saturation of water is increased from the irreducible saturation (the right-hand panel in Fig. 7.3), the relative permeability of air decreases and reaches zero at the residual air saturation. At this point the air phase becomes discontinuous and no flow is possible. Meanwhile, as the water saturation increases, the relative permeability of the water phase increases until it reaches a maximum at the residual air saturation.

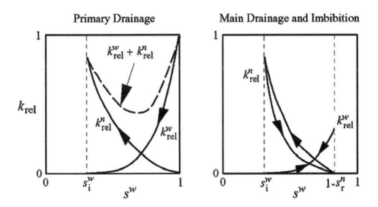

Figure 7.3 Typical relative permeability curves for drainage and imbibition.

Consider now the term in Eq. (7.21) of the form

$$\varepsilon\rho^{w}\frac{DS^{w}}{Dt}.$$

Employing the saturation-pressure relationship of Fig. 7.2 and the chain rule one can write

$$\varepsilon \rho^w \frac{DS^w}{Dt} = \varepsilon \rho^w \frac{\partial S^w}{\partial p_c} \frac{Dp_c}{Dt}. \tag{7.43}$$

Since

$$S^w + S^a = 1 \tag{7.44}$$

then

$$\frac{DS^w}{Dt} = -\frac{DS^a}{Dt} \tag{7.45}$$

which allows us to use the same saturation pressure curve we introduced earlier to obtain

$$\varepsilon \rho^a \frac{DS^a}{Dt} = -\varepsilon \rho^a \frac{\partial S^w}{\partial p_c} \frac{Dp_c}{Dt}. \tag{7.46}$$

Employing Eq. (5.5) on page 87 we can address the term $S^a \varepsilon \frac{D\rho^a}{Dt}$ appearing in Eq. (7.40) as follows:

$$S^w \varepsilon \frac{D\rho^w}{Dt} = S^w \varepsilon^w \rho^w \beta^w \frac{Dp^w}{Dt}. \tag{7.47}$$

Consider now the term in Eq. (7.41) of the form

$$-\frac{S^\alpha \rho^\alpha}{(1-\varepsilon) \rho^s} \frac{D\left[(1-\varepsilon) \rho^s\right]}{Dt}. \tag{7.48}$$

We can write this as

$$\begin{aligned}
&-\frac{S^\alpha \rho^\alpha}{(1-\varepsilon) \rho^s} \frac{D\left[(1-\varepsilon) \rho^s\right]}{Dt} \\
&= -S^\alpha \rho^\alpha \frac{1}{(1-\varepsilon) \rho^s} \left[\frac{\partial\left[(1-\varepsilon) \rho^s\right]}{\partial p^s} \frac{Dp^s}{Dt}\right] \\
&= -S^\alpha \rho^\alpha \alpha^b \frac{Dp^s}{Dt}
\end{aligned} \tag{7.49}$$

where

$$\alpha^b = \frac{1}{(1-\varepsilon) \rho^s} \frac{\partial\left[(1-\varepsilon) \rho^s\right]}{\partial p^s} \tag{7.50}$$

is the *matrix compressibility*.

The term

$$\frac{S^\alpha \rho^\alpha (1-\varepsilon)}{\rho^s} \frac{D\rho^s}{Dt} \tag{7.51}$$

can be written

$$\frac{S^\alpha \rho^\alpha (1-\varepsilon)}{\rho^s} \frac{D\rho^s}{Dt} = \left(\frac{S^\alpha \rho^\alpha (1-\varepsilon)}{\rho^s}\right) \frac{\partial \rho^s}{\partial p^s} \frac{Dp^s}{Dt} \tag{7.52}$$

or

$$\frac{S^\alpha \rho^\alpha (1-\varepsilon)}{\rho^s} \frac{D\rho^s}{Dt} = S^\alpha \rho^\alpha (1-\varepsilon) \beta^s \frac{Dp^s}{Dt} \tag{7.53}$$

where

$$\beta^s \equiv \frac{1}{\rho^s} \frac{\partial \rho^s}{\partial p^s} \tag{7.54}$$

is the *solid compressibility*.

The *divergence of the flux* can be addressed as follows using Eq. (7.42):

$$\boldsymbol{\nabla} \cdot \left(\rho^\alpha \underline{q}^\alpha \right) = -\boldsymbol{\nabla} \cdot \left[\rho^\alpha \left(\frac{k_{r\alpha} \underline{\underline{k}}}{\mu} \left(\nabla p^\alpha + \rho^\alpha \hat{g} \nabla z \right) \right) \right]. \tag{7.55}$$

Left to consider are the interphase flux terms; these terms describe the movement of phase components across their phase boundaries. In the water-air case we are concerned primarily with evaporation and condensation of water and the formation of ice. The terms of interest are

$$-e_{wa}^\alpha - e_{\alpha s}^\alpha + \frac{\rho^\alpha \varepsilon S^\alpha}{\rho^s (1-\varepsilon)} \left(e_{ws}^s + e_{as}^s \right) \quad \text{with } \alpha = w, a. \tag{7.56}$$

In an isothermal system such as we are considering here, there is no mass transfer between phases and these terms vanish.

We now collect the relevant information to create the equation describing unsaturated flow in a porous medium. We begin with the water flow equation

$$-\frac{S^w \rho^w}{(1-\varepsilon)\rho^s} \frac{D[(1-\varepsilon)\rho^s]}{Dt} + \frac{S^w \rho^w (1-\varepsilon)}{\rho^s} \frac{D\rho^s}{Dt} + \varepsilon\rho^w \frac{DS^w}{Dt} + S^w \varepsilon \frac{D\rho^w}{Dt}$$
$$+ \boldsymbol{\nabla} \cdot \left(\rho^w \underline{q}^w \right) = 0. \tag{7.57}$$

Combining this expression with the information in Eqs. (7.22) through (7.27) we obtain

$$-S^w \rho^w \alpha^b \frac{Dp^s}{Dt} + S^w \rho^w (1-\varepsilon) \beta^s \frac{Dp^s}{Dt} + \varepsilon\rho^w \frac{\partial S^w}{\partial p_c} \frac{Dp_c}{Dt} + S^w \varepsilon^w \rho^w \beta^w \frac{Dp^w}{Dt}$$
$$+ \boldsymbol{\nabla} \cdot \left[\rho^\alpha \left(\frac{k_{r\alpha} \underline{\underline{k}}}{\mu} \left(\nabla p^\alpha + \rho^\alpha \hat{g} \nabla z \right) \right) \right] = 0. \tag{7.58}$$

If we make the assumption that grain compressibility and the velocity of the solid grains are small, we can simplify Eq. (7.29) to give

$$-S^w \rho^w \alpha^b \frac{\partial p^s}{\partial t} + \varepsilon\rho^w \frac{\partial S^w}{\partial p_c} \frac{\partial p_c}{\partial t} + S^w \varepsilon^w \rho^w \beta^w \frac{\partial p^w}{\partial t}$$
$$+ \boldsymbol{\nabla} \cdot \left[\rho^\alpha \left(\frac{k_{r\alpha} \underline{\underline{k}}}{\mu} \left(\nabla p^w + \rho^w \hat{g} \nabla z \right) \right) \right] = 0. \tag{7.59}$$

We now assume that the pressure on the solid can be written as the weighted average of the water and air pressures acting on the surface of the grains, that is [17],

$$p^s = \chi p^w + (1-\chi) p^a \tag{7.60}$$

where χ is the Bishop parameter. If we assume that the water pressure is dominant and select $\chi = 1$, we have that

$$p^s = p^w \tag{7.61}$$

and recalling that $p_c = p^w - p^a$, we can rewrite Eq. (7.59) as

$$-S^w \rho^w \alpha^b \frac{\partial p^w}{\partial t} + \varepsilon \rho^w \frac{\partial S^w}{\partial p_c} \frac{\partial (p^w - p^a)}{\partial t} + S^w \varepsilon^w \rho^w \beta^w \frac{\partial p^w}{\partial t}$$
$$+ \nabla \cdot \left[\rho^w \left(\frac{k_{rw} \underline{\underline{k}}}{\mu} \left(\nabla p^w + \rho^w \hat{g} \nabla z \right) \right) \right] = 0. \tag{7.62}$$

A similar development provides the air-phase equation, that is

$$- \left(1 - S^w \right) \rho^a \alpha^b \frac{\partial p^a}{\partial t} + \varepsilon \rho^w \frac{\partial (1 - S^w)}{\partial p_c} \frac{\partial (p^w - p^a)}{\partial t}$$
$$+ \left(1 - S^w \right) \varepsilon^a \rho^a \beta^a \frac{\partial p^a}{\partial t} + \nabla \cdot \left[\rho^a \left(\frac{k_{rw} \underline{\underline{k}}}{\mu} \left(\nabla p^a + \rho^w \hat{g} \nabla z \right) \right) \right] = 0. \tag{7.63}$$

Given the supplementary information contained in the constitutive relationships, all of the state variables appearing in Eqs. (7.32) and (7.36) can be expressed in terms of p^w and p^a. This set of nonlinear equations, coupled through the pressures, must be solved simultaneously.

7.5   THE IMMOBILE AIR UNSATURATED FLOW MODEL

In soil science it is traditional to assume that air, while present, is an immobile phase. When this assumption is made, one is implicitly accepting that:

1. The air phase pressure is essentially constant and atmospheric; and

2. There is no significant air movement.

Given these assumptions, the air-phase equation vanishes and we have only one equation, which is of the form

$$\left[\varepsilon \rho^w \frac{\partial S^w}{\partial p_c} + S^w \varepsilon^w \rho^w \beta^w - S^w \rho^w \alpha^b \right] \frac{\partial p^w}{\partial t}$$
$$+ \nabla \cdot \left[\rho^w \left(\frac{k_{rw} \underline{\underline{k}}}{\mu} \cdot \left(\nabla p^w + \rho^w \hat{g} \nabla z \right) \right) \right] = 0. \tag{7.64}$$

Equation (7.64) is an extension of the Richards equation, which in its original form is

$$\frac{\partial \varepsilon S^w}{\partial t} = \frac{\partial}{\partial z} \left[K \left(S^w \right) \left(\frac{\partial \left(p^w / (\rho^w g) \right)}{\partial z} + 1 \right) \right] \tag{7.65}$$

where $K = K(S^w)$ is the hydraulic conductivity, defined in this case as $\frac{k_{rw} k}{\mu}$. Note that the system of equations has been reduced to one non-linear equation in the pressure p^w.

7.6 BOUNDARY CONDITIONS

The boundary conditions for the saturated flow model presented in Section 7.3 can be first, second, or third type defined on the boundary $\partial\Omega$ of region Ω. In the case of the pressure-based formulation the first-type (also called Dirichlet) condition is a specified pressure, that is

$$p\left(\underline{x}, t\right) = p_o\left(\underline{x}, t\right), \qquad \underline{x} \in \partial\Omega_1. \tag{7.66}$$

The second-type (also known as Neumann) condition is usually presented as a flux, that is,

$$\underline{q}_o \cdot \underline{n} = -\frac{\underline{\underline{k}}}{\mu} \cdot \left(\nabla p + \rho\hat{g}\nabla z\right) \cdot \underline{n} \tag{7.67}$$

or, rearranging

$$\nabla p\left(\underline{x}, p\right) \cdot \underline{n} = -\mu\underline{\underline{k}}^{-1}\underline{q}_o\left(\underline{x}, t\right) \cdot \underline{n} - \rho\left(\underline{x}, t\right)\hat{g}\nabla z \cdot \underline{n}, \quad \underline{x} \in \partial\Omega_2 \tag{7.68}$$

where $\partial\Omega = \partial\Omega_1 + \partial\Omega_2 + \partial\Omega_3$. Note that for the common condition of no flow across the boundary,

$$\nabla p\left(\underline{x}, p\right) \cdot \underline{n} = -\rho\left(\underline{x}, t\right)\hat{g}\nabla z \cdot \underline{n}. \tag{7.69}$$

The third-type condition is seldom used in the pressure formulation and will not be presented.

The alternative formulation for the saturated flow equations uses the concept of hydraulic head h^w as the unknown state variable for which a solution is sought. The case of constant head is a first-type or Dirichlet condition of the form

$$h^w\left(\underline{x}, t\right) = h_0^w\left(\underline{x}, t\right). \tag{7.70}$$

Practically speaking, this condition applies to the situation where the head value at a location on the boundary is known, such as would be the case where a portion of the boundary $\partial\Omega_1$ is in contact with a surface water body.

Since the water table is defined as a surface along which the pressure is defined as atmospheric, along this boundary a constant-pressure atmospheric condition or, equivalently, a head value equal to the elevation of the water table may be applied.

The Neumann or second-type condition when specified using head is of the form

$$\underline{q}_o \cdot \underline{n} = -\underline{\underline{K}} \cdot \nabla h^w \cdot \underline{n} \tag{7.71}$$

which upon rearrangement gives

$$\nabla h^w \cdot \underline{n} = -\underline{\underline{K}}^{-1} \cdot \underline{q}_o \cdot \underline{n} \tag{7.72}$$

where $\underline{\underline{K}}^{-1}$ is the inverse of the matrix $\underline{\underline{K}}$. In practical applications the Neumann condition identifies flux conditions across a portion of the boundary $\partial\Omega$, that is $\partial\Omega_2$. As noted earlier, the most common use of this condition is to define a boundary across which there is no flow; for example, where an aquifer is in contact with an imperme-able geological formation. However, there are also hydrodynamic boundaries where there is no flow. Consider, for example, the groundwater flow conditions beneath the crest of a water table where the head drops off symmetrically in all directions. At this point, that is, at the crest of the water table surface, there is no horizontal gradient in head or pressure. Thus the vertical line beneath this point constitutes a no-flow condition in the horizontal plane. Keep in mind, however, that if conditions change and the water table at this point is disturbed, the boundary condition is no longer defined by symmetry.

The third-type (also called Robbin's) condition is of the form

$$\gamma \nabla h^w \cdot \underline{n} + h^w = f_0 (\underline{x}, t) \tag{7.73}$$

where $f_0 (\underline{x}, t)$ is a known function. It is a linear combination of the Neumann and Dirichlet conditions noted above. One common application of this type of boundary condition in saturated flow is of the form

$$\frac{\partial h}{\partial z} = -\kappa (h^w - h^\sigma) \tag{7.74}$$

where κ is a parameter describing the ability of water to flow through a layer separating the aquifer where the head is h^w and a surface water body with head h^σ. Thus this condition can be thought of as describing leakage into the aquifer.

In the case of unsaturated flow using a single equation formulation, such as described in Section 7.5, the conditions defined above for the pressure formulation hold. Since the air phase in this case is assumed to be a constant, one can also specify the saturation $S^w (p_c)$, since by specifying S^w one also specifies p_c through the saturation-pressure relationship and since p^a is constant one also specifies p^w, the boundary condition sought . Similarly, by specifying the flux entering the boundary, one defines the gradient in pressure. However, there is a complication because, for the case of unsaturated flow, the condition is of the form

$$\nabla p^w (\underline{x}, t) \cdot \underline{n} = -\frac{\mu \underline{\underline{k}}^{-1}}{k_{rw} (p_c)} \underline{q}_o (\underline{x}, t) \cdot \underline{n} - \rho^w (\underline{x}, t) \hat{g} \nabla z \cdot \underline{n}, \quad \underline{x} \in \partial \Omega_2 \tag{7.75}$$

which has the relative permeability $k_{rw} (p_c)$ as part of the boundary condition. Thus this is a non-linear boundary condition and is more complex to address numerically than the corresponding linear boundary condition.

In the case of multiphase flow, that is, when the air and water phases are dynamic, the specification of a Dirichlet, constant-pressure condition must be considered with some care. Recall once again that the capillary pressure links the air and water pressure through the relationship $p_c = p_w - p_n$. Moreover, the saturation is dependent on the capillary pressure, that is, $S^w = S^w (p_c)$. Thus the two pressures and the saturation are linked and cannot be defined independently. In other words, if the saturation at the boundary is known, the pressures used on the boundary must be self-consistent with this saturation, and further, the two fluid pressures must be consistent with the capillary pressure.

7.7 SUMMARY

The objective of this chapter is to present the formulation of the equations that describe the flow and transport of multiphase fluids in a porous medium. The point balance equations for a single-phase system are derived from the basic model used earlier in the book. From the point balance equations for a species i in a phase α, the porous medium balance equations are derived. It is shown that by summing the species balance equations, a fluid balance equation can be obtained. The resulting multiphase formulation is subsequently modified to consider the special cases of an air-water system and an air-water-gas system; the former is identified with environmental engineering and the latter with petroleum engineering.

EXERCISES

7.1   The relative permeability curves in Fig. 7.3 are a nonlinear function of the saturation of water. Intuitively, one might imagine that the relative permeability would depend directly on the water saturation since it describes the amount of air present in the porous medium that is blocking the pores. Present an argument based upon porous-flow-physics principles that could explain why these curves are not linear.

7.2   The following relationship (Eq. 7.33)

$$h^w = \int_{p_{ref}}^{p^w} \frac{dp^{w'}}{\rho\left(p^{w'}\right)g} - \frac{g \cdot k}{g}\left(z - z_{ref}\right) \tag{7.76}$$

is used on page 136 to develop the the air-water equation in terms of head h^w rather than pressure. In the course of that development, it was argued that this relationship is applicable only when the density depends only on pressure. Explain the rationale behind this constraint. The total head h_w is composed of the elevation head plus the pressure head. Which terms in the integrand represent these two head values?

7.3   You design a drainage system and place your drain at a depth of 12 ft. below the surface. You want to know exactly how far down the soil moisture is such that the saturation is at 0.25 (25%); i.e., point B in Fig. 7.4. The goal is to determine the depth L exactly so that you can plant on surface A crops with the correct root depth such that they will have optimal moisture. The challenge is to determine the depth L at which you will encounter the 25% moisture content. The drain will automatically keep the water table (zero pressure surface) at location midway through the drain, as shown at location C. Given that the diameter of the drain pipe is 6 inches, determine L. You will need the information provided in Fig. 7.5. Assume that there is no net infiltration from the surface (rainfall) so that the water column is stationary (not flowing). Hint: It is helpful to think in terms of head.

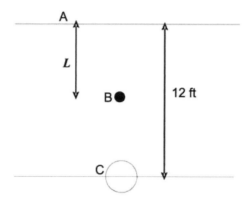

Figure 7.4 Cross section of drainage plan.

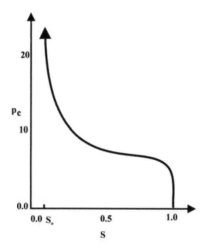

Figure 7.5 Saturation-capillary pressure curve. The horizontal axis is saturation (as a fraction) and the vertical axis is capillary pressure (negative water pressure) in feet of water.

7.4 In Eq. 7.20 on page 134 we expressed the flow equation as

$$\frac{D\left(\varepsilon^w \rho^w\right)}{Dt} - \frac{\varepsilon^w \rho^w}{\varepsilon^s \rho^s}\frac{D\left(\varepsilon^s \rho^s\right)}{Dt} + \nabla \cdot \left(\rho^w \underline{q}\right) - e_{ws}^w + \frac{\varepsilon^w \rho^w}{\varepsilon^s \rho^s}e_{sw}^s = 0 \qquad (7.77)$$

where $\frac{D(\cdot)}{Dt}$ is the substantial derivative expressed in terms of the grain velocity. With this definition in mind, explain the physical meaning of the first two terms of Eq. (7.77).

7.5 In Eq. (7.6) on page 132 we introduce the term $e_\beta^{i\alpha}$ as the mass (per unit total volume of the system) of species i that goes from phase α to phase β, across the interface that separates these two phases. The movement across the interface can be due to three phenomena: the movement of the species when the interface is stationary, the movement of the interface when the species are stationary, and a combination of the above. Explain under what physical situations you would expect these three phenomena to take place.

7.6 Figure 7.6 is a graph of the total head versus elevation in a sand column. Assume that the reference location, b, is zero for both. At what elevation, A, B, C, or D, can the sand in this column begin to be partially saturated? Explain why.

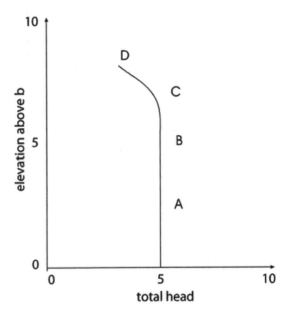

Figure 7.6 Plot of head versus elevation.

REFERENCES

1. Allen, M.B., G.A. Behie and J.A. Trangenstein, *Multiphase Flow in Porous Media: Mechanics and Numerics*, Lecture Notes in Engineering, Vol.34, Eds. C.A. Brebbia and S.A. Orszag, Springer-Verlag, Berlin, 1988

2. Aziz, K. and A. Settari, *Petroleum Reservoir Simulation*, Applied Science Publishers, London, 1979.

3. Chen, Z., G. Huan, and Y. Ma, *Computational Methods for Multiphase Flows in Porous Media*, SIAM, Philadelphia, 2006.

4. Cheng, A.H.-D., *Multilayered Aquifer Systems: Fundamentals and Applications*, Marcel Dekker, New York, 2000.

5. De Marsily, G., *Quantitative Hydrogeology: Groundwater Hydrology for Engineers*, Academic Press, New York, 1986.

6. Ewing, W.M., W.S. Jardewsky, and F. Press, *Elastic Waves in Layered Media*, McGraw-Hill, New York, 1957.

7. Fanchi, J.R., *Shared Earth Modeling*, Butterworth Heinemann and Elsevier Science, New York, 2002.

8. Hellwig, G., *Partial Differential Equations: An Introduction*, Blaisdell, New York, 1964.

9. Helmig, R., *Multiphase Flow and Transport Processes in the Subsurface: A Contribution to the Modeling of Hydrosystems*, Springer-Verlag, Berlin, 1997.

10. Herrera, I., A. Galindo, and R. Camacho, Shock modelling in variable bubble point problems of petroleum engineering, *Computational Modelling of Free and Moving Boundary Problems*, Vol. 1: Fluid Flow, Eds. L.C. Wrobel and C.A. Brebbia, CMP Books, San Francisco, CA, 399-415, 1991.

11. Herrera, I., R. Camacho, and A. Galindo, Shock modelling in petroleum engineering, Chapter 7 of *Computational Methods for Moving Boundary Problems in Heat and Fluid Flow*, Eds. L.C. Wrobel and C.A. Brebbia, CMP Books, San Fracisco, CA, 143-170, 1993.

12. Herrera, I. and L. Chargoy, Shocks in solution gas-drive reservoirs., SPE Thirteenth Symposium in Reservoir Simulation, San Antonio, TX, Paper SPE 029136, 429-440, 1995.

13. Herrera, I., Shocks and bifurcations in black-oil models, *SPE J,* 1(1), 51-58, 1996.

14. Mattax, C.C. and R.L. Dalton, *Reservoir Simulation*, SPE Monograph 13, Society of Petroleum Engineers, Richarson, TX, 1990.

15. Muskat, M., *Physical Principles of Oil Production*, McGraw-Hill, New York, 1949.

16. Peaceman, D.W., *Fundamentals of Numerical Reservoir Simulation*, Elsevier, Amsterdam, 1977.

17. Pinder, G.F. and W.G. Gray, *Essentials of Multiphase Flow and Transport in Porous Media*, Wiley, Hoboken, NJ, 2008.

18. Szilas, A. P., *Production and Transport of Oil and Gas*, second completely revised edition, Part A: Flow Mechanics and Production, Developments in Petroleum Science, Vol.18A, Elsevier, Amsterdam, 1985.

19. Szilas, A.P., *Production and transport of oil and gas*, second completely revised edition, Part B: Gathering and Transport, Developments in Petroleum Science, Vol.18B, Elsevier, Amsterdam, 1986.

20. Thomas, G.W., *Principles of Hydrocarbon Simulation*, International Human Resources Development Corporation, Boston, 1982.

CHAPTER 8

ENHANCED OIL RECOVERY

8.1 BACKGROUND ON OIL PRODUCTION AND RESERVOIR MODELING

Generally, there are three stages of recovery that are identified in the production life of a petroleum reservoir: primary, secondary, and tertiary. *Primary recovery* refers to the production that is obtained using the energy inherent in the reservoir due to gas under pressure or a natural *water drive*. At a very early stage the reservoir essentially contains a single fluid such as gas or oil (the presence of water can usually be neglected) and often the pressure is so high that the oil or gas is produced without any pumping of the wells. *Primary recovery* ends when the oil field and the atmosphere reach pressure equilibrium. The total recovery obtained at this stage is usually around 12-15% of the hydrocarbons contained in the reservoir (OIIP: oil initially in place).

The technique of *waterflooding* used to be considered as an *enhanced oil recovery* method, but nowadays *secondary recovery* usually refers to waterflooding. In this approach water is injected into some wells *(injection wells)* to maintain the field pressure and flow rates, while oil is produced through other wells *(production wells)*. In secondary recovery, if the oil phase is above the *bubble point,* the flow is *two-phase immiscible,* with water in one phase and oil in the other. In such a case, there

Mathematical Modeling in Science and Engineering: An Axiomatic Approach.
By Ismael Herrera and George F. Pinder Copyright © 2012 John Wiley & Sons, Inc.

is no mass exchange between the phases. When the pressure drops below the bubble point the hydrocarbon component of the system separates into two phases: oil and gas. This is the case to which the *black-oil* model applies; in it, the oil and gas phases exchange mass while the water phase does not. *Secondary recovery* yields an additional 15-20% of the OIIP.

After secondary recovery has been completed, 50% or more of the hydrocarbons often remains in the reservoir. The more advanced techniques that have been developed for recovering such a valuable volume of hydrocarbons are known as *tertiary recovery techniques* or by the more generic term *enhanced oil recovery (EOR)*.

The terms *primary*, *secondary*, and *tertiary* may be confusing. For example, water injection (a secondary recovery strategy) is often implemented from the start in the North Sea, and cyclic steam injection is also often applied from the start in heavy oil reservoirs. Actually, EOR methods are employed to obtain additional yields from a reservoir either after secondary recovery procedures have been applied or to treat non-conventional fields, which, due to their difficult characteristics, require advanced methods from the start. In this respect, the term EOR (also called *improved oil recovery; IOR*) is more adequate. A suitable definition of EOR is: *Enhanced oil recovery processes are those methods that use external sources of materials and energy to recover oil from a reservoir that cannot be produced economically by conventional means*

The most important EOR methods can be grouped as follows:

1. *Waterflooding* (conventional, water-alternating-gas (WAG), polymer flooding);

2. *Miscible gas injection*: hydrocarbon gas, CO_2, nitrogen, flue gas;

3. *Chemical injection*: polymer/surfactant, caustic and micellar/polymer flooding; and

4. *Thermal oil recovery*: cyclic steam injection, steam flooding, hot-water drive, in-situ combustion.

At present a large part of the oil reserves of the world are located in mature oil fields whose production is declining and for which the only possibility for expanding their yield is by application of EOR techniques. Another large fraction of such oil reserves are non-conventional oil fields that are very difficult to exploit, either because of the characteristics of their hydrocarbons, such as very high viscosity, or because of the characteristics of the soils and rocks in which they are contained. In both cases, such reservoirs can only be exploited by applying EOR methods. Therefore, today, EOR is an important strategy for sustaining the oil supply of the world.

On the other hand, mathematical and computational modeling of oil reservoirs is fundamental for the development and application of EOR techniques. This is because computational modeling supplies an important methodology for predicting and understanding the behavior of a reservoir when it is subjected to the complicated and varied processes that constitute EOR methods. In addition, oil-reservoir modeling is essential for defining suitable strategies for maximizing oil recovery, and it facilitates predicting the future performance of oil fields, including overall yield.

8.2   PROCESSES TO BE MODELED

In primary production, the processes to be modeled are the motion of one phase or of two phases at most, without mass exchange between the phases. In secondary production generally, the motion of a three-phase fluid system has to be modeled: the phases being water, oil, and gas. Mass exchange between the oil and gas phases must be included. The standard computational model to mimic such a system is technically known as the *black-oil model*.

Another process to be modeled is *multiphase-multispecies transport* in two modalities. One is in isothermal conditions and the other is in non-isothermal conditions. The main difference between one and the other formulation is that in the second case, energy balances must be included. The particular study of in-situ combustion is further complicated by the fact that sharp fronts (shocks) may need to be included.

8.3   UNIFIED FORMULATION OF EOR MODELS

The axiomatic manner of deriving the basic mathematical models of multiphase systems was explained in Chapter 3. For any multiphase system such a model is defined by a family of extensive properties $\left\{E^1, ..., E^N\right\}$, which has an associated family of intensive properties $\left\{\psi^1, ..., \psi^N\right\}$. Here, N is the cardinality of such families (i.e., the number of members they contain). Then the basic mathematical model consists of the following balance conditions:

$$\left.\begin{array}{c} \dfrac{\partial \psi^\alpha}{\partial t} + \nabla \cdot (\psi^\alpha \underline{v}^\alpha) = g^\alpha + \nabla \cdot \underline{\tau}^\alpha \ \text{ in } \ \Omega - \Sigma \\[2mm] \llbracket \psi^\alpha (\underline{v}^\alpha - \underline{v}_\Sigma) - \underline{\tau}^\alpha \rrbracket \cdot \underline{n} = 0 \ \text{ on } \ \Sigma \end{array}\right\}, \quad \alpha = 1, ..., N. \qquad (8.1)$$

It is also useful to recall the global balance equations, Eq. (1.25):

$$\frac{dE^\alpha}{dt}(t) = \int_{B(t)} g^\alpha(\underline{x}, t)\, d\underline{x} + \int_{\partial B(t)} \underline{\tau}^\alpha(\underline{x}, t) \cdot \underline{n}(\underline{x}, t)\, d\underline{x}, \ \ \alpha = 1, ..., N. \ \ (8.2)$$

To use this unified procedure in developing the EOR models we have to identify, for each of them, a family of extensive properties and apply to each member of such a family the balance conditions in terms of the associated intensive property. This yields a system of N differential equations and N jump conditions. We recall that in the case of the multiphase systems we will be dealing with, each of the extensive properties is associated with one and only one of the phases of the system, and the balance conditions of Eq. (8.1) must be applied with the particle velocity of such a phase. In each of the cases to be treated, after having obtained the basic mathematical model, we shall incorporate the knowledge about the system through suitable constitutive equations.

8.4   THE BLACK-OIL MODEL

The basic assumptions on which this model is built are:

1. There are three phases: water, oil (a liquid phase of hydrocarbons), and gas (a gas phase of hydrocarbons);

2. In the oil phase there are two components: non-volatile oil and volatile oil (dissolved gas);

3. Each of the other two phases is made of only one component;

4. The processes occurring in the reservoir are isothermal, so the balance of energy is not incorporated in the analysis;

5. Between the oil phase and the gas phase there is mass exchange of volatile oil, since gas may dissolve in the oil phase and vice versa. Except for this, no mass exchange occurs between the different phases;

6. There is no adsorption by the solid matrix nor chemical reaction between the different components of the system; and

7. In each of the phases, diffusion-dispersion of the components is neglected.

The family of extensive properties on which the black-oil model is based consists of four masses: namely, the masses of the water and gas phases, and the masses of the two components contained in the oil phase: non-volatile oil and dissolved gas. These masses will be denoted by M^w, M^g, M^{Oo}, and M^{Go}, respectively.

In a straightforward manner, the integral expressions of these properties can be written as

$$
\begin{aligned}
M^w(t) &= \int_{B(t)} \varepsilon S_w \rho^w \, dx \\
M^{Oo}(t) &= \int_{B(t)} \varepsilon S_o \rho^{Oo} \, dx \\
M^{Go}(t) &= \int_{B(t)} \varepsilon S_o \rho^{Go} \, dx \\
M^g(t) &= \int_{B(t)} \varepsilon S_g \rho^g \, dx.
\end{aligned}
\tag{8.3}
$$

In view of Eq. (8.3), it is clear that the associated family of intensive properties consists of

$$
\begin{aligned}
\psi^w &= \varepsilon S_w \rho^w \\
\psi^{Oo} &= \varepsilon S_o \rho^{Oo} \\
\psi^{Go} &= \varepsilon S_o \rho^{Go} \\
\psi^g &= \varepsilon S_g \rho^g.
\end{aligned}
\tag{8.4}
$$

The density of a phase α, namely ρ^α, is understood to be the total mass of that phase per unit volume of the phase, while $\rho^{\beta\alpha}$ is the mass of component β in phase α per unit volume of the phase. Thus, for example, if we use ρ^o for the density of the oil phase, then

$$
\rho^o = \rho^{Oo} + \rho^{Go}.
\tag{8.5}
$$

For the black-oil model, the global balance equations of Eq. (8.2) are:

$$
\left|
\begin{aligned}
\frac{dM^w}{dt}(t) &= \int_{B(t)} g^w(\underline{x},t)\,dx + \int_{\partial B(t)} \underline{\tau}^w(\underline{x},t) \cdot \underline{n}(\underline{x},t)\,dx \\[2mm]
\frac{dM^{Oo}}{dt}(t) &= \int_{B(t)} g^{Oo}(\underline{x},t)\,dx + \int_{\partial B(t)} \underline{\tau}^{Oo}(\underline{x},t) \cdot \underline{n}(\underline{x},t)\,dx \\[2mm]
\frac{dM^{Go}}{dt}(t) &= \int_{B(t)} g^{Go}(\underline{x},t)\,dx + \int_{\partial B(t)} \underline{\tau}^{Go}(\underline{x},t) \cdot \underline{n}(\underline{x},t)\,dx \\[2mm]
\frac{dM^g}{dt}(t) &= \int_{B(t)} g^g(\underline{x},t)\,dx + \int_{\partial B(t)} \underline{\tau}^g(\underline{x},t) \cdot \underline{n}(\underline{x},t)\,dx.
\end{aligned}
\right.
\tag{8.6}
$$

We now incorporate the general assumptions on which the black-oil model is based. First, we notice that $\underline{\tau}^\alpha = 0$, $\alpha = w, Oo, Go, g$, since the diffusion-dispersive processes in the phases have been neglected. As for g^w, the mass of the water inside the system is conserved; however, generally, water together with the other fluid components may be extracted from or injected into the reservoir through wells, and in large-scale studies when this happens the injection-extraction procedure is incorporated taking $g^w \neq 0$. Something similar happens with g^{Oo}, g^{Go}, and g^g. However, in the case of g^{Go} there is an additional contribution to it because there is mass exchange between the oil and gas phases. Taking this into account, we separate g^{Go} into two terms, one accounting for the injection-extraction processes, g_E^{Go}, and the other for exchange between the phases g_g^{Go}. A similar analysis applies to the source g^g, which can also be separated into two parts. Thus, we write

$$
g^{Go} \equiv g_E^{Go} + g_g^{Go} \quad \text{and} \quad g^g \equiv g_E^g + g_{Go}^g.
\tag{8.7}
$$

It should be observed that conservation of mass of the volatile oil imposes the condition that

$$
g_g^{Go} + g_{Go}^g = 0.
\tag{8.8}
$$

Now, a straightforward application of Eq. (8.1) yields the following differential equations:

$$
\begin{aligned}
\frac{\partial \varepsilon S_w \rho^w}{\partial t} + \nabla \cdot (\varepsilon S_w \rho^w \underline{v}^w) &= g^w \\[2mm]
\frac{\partial \varepsilon S_o \rho^{Oo}}{\partial t} + \nabla \cdot (\varepsilon S_o \rho^{Oo} \underline{v}^o) &= g^{Oo} \\[2mm]
\frac{\partial \varepsilon S_o \rho^{Go} \rho^o}{\partial t} + \nabla \cdot (\varepsilon S_o \rho^{Go} \underline{v}^o) &= g_E^{Go} + g_g^{Go} \\[2mm]
\frac{\partial \varepsilon S_g \rho^g}{\partial t} + \nabla \cdot (\varepsilon S_g \rho^g \underline{v}^g) &= g_E^g + g_{Go}^g
\end{aligned}
\tag{8.9}
$$

as well as the jump conditions:

$$\llbracket \varepsilon S_w \rho^w (\underline{v}^w - \underline{v}_\Sigma) \rrbracket \cdot \underline{n} = 0$$

$$\llbracket \varepsilon S_o \rho^{Oo} (\underline{v}^o - \underline{v}_\Sigma) \rrbracket \cdot \underline{n} = 0$$

$$\llbracket \varepsilon S_o \rho^{Go} (\underline{v}^o - \underline{v}_\Sigma) \rrbracket \cdot \underline{n} = 0 \qquad (8.10)$$

$$\llbracket \varepsilon S_g \rho^g (\underline{v}^g - \underline{v}_\Sigma) \rrbracket \cdot \underline{n} = 0$$

which must be fulfilled at the shocks, Σ, where discontinuities of the intensive properties occur. In petroleum engineering some of the most extensively studied shocks are those of the Buckley-Leverett type [2]. Using the axiomatic formulation, shocks that occur in variable-bubble-point reservoirs were reported and analyzed by Herrera et al. [[9] , [8], and [6]].

However, to simplify our presentation a little, in what follows we discuss the governing differential equations exclusively. To obtain them in a form that is most commonly used in practice, we first introduce the multiphase form of the Darcy velocities, which are defined for each of the phases by

$$\underline{u}_\alpha \equiv \varepsilon S_\alpha \underline{v}^\alpha, \quad \alpha = w, o, g \qquad (8.11)$$

and then we sum the last two equations of Eq. (8.9), making use of Eq. (8.8) to obtain

$$\frac{\partial \varepsilon \rho^w S_w}{\partial t} + \nabla \cdot (\rho^w \underline{u}_w) = g^w$$

$$\frac{\partial \varepsilon \rho^{Oo} S_o}{\partial t} + \nabla \cdot (\rho^{Oo} \underline{u}_o) = g^{Oo} \qquad (8.12)$$

$$\frac{\partial \varepsilon (\rho^{Go} S_o + \rho^g S_g)}{\partial t} + \nabla \cdot (\rho^{Go} \underline{u}_o + \rho^g \underline{u}_g) = g^G.$$

Here $g^G \equiv g_E^{Oo} + g_E^{Go}$ is the total volatile oil that enters or leaves the reservoir due to *injection-extraction* processes.

Standard versions of the black-oil model incorporate at least the following *constitutive equations*:

1. Darcy's law for each phase:

$$\underline{u}_\alpha = -\frac{\underline{\underline{k}}_\alpha}{\mu_\alpha} (\nabla p^\alpha - \rho^\alpha \hat{g} \nabla z), \quad \alpha = w, o, g. \qquad (8.13)$$

This is the generalization to multiphase flow of Darcy's law, introduced in Chapter 5. In Eq. (8.13), $\underline{\underline{k}}_\alpha$, μ_α, and $\hat{g}$ are the effective permeability, the viscosity of phase α, and the magnitude of the gravitational acceleration, respectively. The effective permeability is usually expressed as

$$\underline{\underline{k}}_\alpha = k_{r\alpha} \underline{\underline{k}}, \quad \alpha = w, o, g. \qquad (8.14)$$

Here $\underline{k}$ and $k_{r\alpha}$ are the absolute and relative permeabilities, respectively.

2. *The saturation identity:* this is stated as

$$S_w + S_o + S_g = 1. \tag{8.15}$$

This relationship is due to the assumption, usually made in petroleum reservoir mechanics, that the porous medium is saturated (that is, the void space of the solid matrix is full of fluids).

3. *The gas oil ratio relation:* this corresponds to writing

$$\rho^{Go} = R_s \rho^{Oo}. \tag{8.16}$$

Here, R_s is a function of the oil pressure and possibly of the temperature. However, the black-oil model is an isothermal model and, consequently, the temperature is a defined datum and not a state variable to be determined.

4. *The capillary pressure relations:*

$$p^o - p^w = p^{cow} \text{ and } p^g - p^o = p^{cgo}. \tag{8.17}$$

This relation has to be satisfied because the pressures in the different phases are not the same and their differences are due to the capillary forces that act at the interfaces that separate them. Generally, the capillary pressures, p^{cow} and p^{cgo}, are functions of the saturations, but the form of such functional dependence in turn depends on many properties of the reservoirs and the fluids contained in them, which are determined experimentally. The capillary pressures are usually assumed to take the general form

$$p^{cow} = p^{cow}(S_w) \text{ and } p^{cgo} = p^{cg\phi}(S_g). \tag{8.18}$$

For the black-oil model, it is often convenient to work with the balance equations on *standard volumes* (these are the volumes at *standard conditions*) instead of the balance equations in mass. When this is done, Eq. (8.12) can be written using the subscript s to designate standard conditions as

$$\frac{\partial}{\partial t}\left(\frac{\varepsilon S_w}{B_w}\right) - \nabla \cdot \left(\underline{\underline{T}}_w \nabla \Phi_w\right) = \frac{q_{Ws}}{B_w}$$

$$\frac{\partial}{\partial t}\left(\frac{\varepsilon S_o}{B_o}\right) - \nabla \cdot \left(\underline{\underline{T}}_o \nabla \Phi_o\right) = \frac{q_{Os}}{B_o} \tag{8.19}$$

$$\frac{\partial}{\partial t}\left\{\varepsilon\left(\frac{S_g}{B_g} + \frac{R_{so}S_o}{B_o}\right)\right\} - \nabla \cdot$$

$$\left(\underline{\underline{T}}_g \nabla \Phi_g + R_{so}\underline{\underline{T}}_o \nabla \Phi_o\right) = \frac{q_{Gs}}{B_g} + \frac{R_{so}q_{Os}}{B_o}.$$

Here

$$B_o \equiv \frac{\rho^{os}}{\rho^o}, \quad B_g \equiv \frac{\rho^{gs}}{\rho^g}, \quad B_w \equiv \frac{\rho^{ws}}{\rho^w},$$

$$\underline{\underline{T}}_\alpha \equiv \frac{k_{r\alpha}}{B_\varepsilon}\underline{\underline{k}}, \quad \Phi_\alpha \equiv p_\alpha - \rho_\alpha \widehat{g} z, \quad \alpha = w, o, g, \quad (8.20)$$

$$R_{so}(p, T) \equiv \frac{\rho^{os} \, \omega^{go} \rho^o}{\rho^{gs} \, \omega^{oo} \rho^o}.$$

Furthermore,

$$q_{Ws} = \frac{g^w}{\rho^{ws}} B_w$$

$$q_{Os} = \frac{g^o}{\rho^{os}} B_o \qquad (8.21)$$

$$q_{Gs} = \frac{g^g}{\rho^{gs}} B_g.$$

As for nomenclature, B_α are the *formation volume factors*, $\underline{\underline{T}}_\alpha$ are the *transmissivities*, and Φ_α are the *fluid potentials*.

8.5 THE COMPOSITIONAL MODEL

In this class of models, known as *compositional models*, a finite number of hydrocarbon components are used to represent the composition of the reservoir fluids. The number of phases considered is at most three and there may be exchange of mass between different phases. For simplicity, we present here the isothermal version of the compositional model, so as in the black-oil model, the processes are assumed to be isothermal. To achieve sufficient generality, the compositional model will be presented for the case for any number of components (N_C) and they may be present in both the oil and gas phases. As shown here, setting up the governing differential equations is an easy task, albeit obtaining their numerical and computational solutions may be very complex and computationally demanding.

The basic assumptions on which this model is built are:

1. There are three phases: water, oil (that is, a liquid phase of hydrocarbons), and gas (that is a gas phase of hydrocarbons);

2. In both the oil and gas phases there are N_C components;

3. The water phase is made of only one component (water);

4. The processes occurring in the reservoir are essentially isothermal, so the balance of energy is not incorporated in the analysis;

5. Between the oil phase and the gas phase there may be mass exchange of each of the components;

6. There is no mass exchange between the water phase and the other phases;

7. There is no adsorption by the solid matrix, nor chemical reaction between the various components of the system; and

8. In each of the phases, diffusion-dispersion of the components is neglected.

The family of extensive properties on which the compositional model is based is made of $2N_C + 1$ masses: namely, the masses of the water phase and those of the N_C components contained in the oil phase as well as those of the N_C components contained in the gas phase. In our notation M^w will be the mass of the water phase, while M^{io} and M^{ig} will be the masses of component i in the oil and gas phases, respectively.

In a straightforward manner, the integral expressions of these masses are

$$\left. \begin{array}{l} M^w(t) = \int_{B(t)} \varepsilon S_w \rho^w \, dx \\ M^{io}(t) = \int_{B(t)} \varepsilon S_o \rho^{io} \, dx \\ M^{ig}(t) = \int_{B(t)} \varepsilon S_g \rho^{ig} \, dx \end{array} \right\} \quad i = 1, ..., N_C. \tag{8.22}$$

In view of Eq. (8.22), it is clear that the associated family of intensive properties includes

$$\left. \begin{array}{l} \psi^w = \varepsilon S_w \rho^w \\ \psi^{io} = \varepsilon S_o \rho^{io} \\ \psi^{ig} = \varepsilon S_g \rho^{ig} \end{array} \right\}, \quad i = 1, ..., N_C. \tag{8.23}$$

For the compositional model, the global balance equations of Eq. (8.21) are

$$\left. \begin{array}{l} \dfrac{dM^w}{dt}(t) = \int_{B(t)} g^w(\underline{x}, t) \, dx + \int_{\partial B(t)} \underline{\tau}^w(\underline{x}, t) \cdot \underline{n}(\underline{x}, t) \, dx \\ \dfrac{dM^{io}}{dt}(t) = \int_{B(t)} g^{io}(\underline{x}, t) \, dx + \int_{\partial B(t)} \underline{\tau}^{io}(\underline{x}, t) \cdot \underline{n}(\underline{x}, t) \, dx \\ \dfrac{dM^{ig}}{dt}(t) = \int_{B(t)} g^{ig}(\underline{x}, t) \, dx + \int_{\partial B(t)} \underline{\tau}^{ig}(\underline{x}, t) \cdot \underline{n}(\underline{x}, t) \, dx \end{array} \right\}, \quad i = 1, ..., N_C. \tag{8.24}$$

First, we notice that $\underline{\tau}^\beta = 0$, $\beta = w, io, ig$, since the diffusion-dispersive processes in the phases have been neglected. Analysis of the other terms is very similar to that of the black-oil model. The mass of the water inside the system is conserved; the injection-extraction through wells is accounted for by taking $g^w \neq 0$. In the case of g^{io} and g^{ig} there is an additional contribution to them because there is mass exchange between the oil and gas phases. Taking this into account, we separate each $g^{i\alpha}$ into two terms, one accounting for the injection-extraction, $g_E^{i\alpha}$, and the other for exchange between the phases $g_{i\beta}^{i\alpha}$, $\alpha \neq \beta$. Thus, we write

$$g^{i\alpha} \equiv g_E^{i\alpha} + g_{i\beta}^{i\alpha}. \tag{8.25}$$

Mass-conservation of each one of the hydrocarbon components implies that

$$g_{ig}^{io} + g_{io}^{ig} = 0, \quad i = 1, ..., N_C. \tag{8.26}$$

Application of Eq. (8.1), together with the definition of the Darcy velocities given in Eq. (8.11), yields

$$\left. \begin{array}{l} \dfrac{\partial \varepsilon S_w \rho^w}{\partial t} + \nabla \cdot \left(\rho^w \underline{u}_w \right) = g^w \\[3mm] \dfrac{\partial \varepsilon S_o \rho^{io}}{\partial t} + \nabla \cdot \left(\rho^{io} \underline{u}_o \right) = g_E^{io} + g_{ig}^{io} \\[3mm] \dfrac{\partial \varepsilon S_g \rho^{ig}}{\partial t} + \nabla \cdot \left(\rho^{ig} \underline{u}_g \right) = g_E^{ig} + g_{io}^{ig} \end{array} \right\}, \quad i = 1, ..., N_C \tag{8.27}$$

as well as the jump conditions

$$\left. \begin{array}{l} [\![\rho^w \left(\underline{u}_w - \underline{u}_\Sigma^w \right)]\!] \cdot \underline{n} = 0 \\[2mm] [\![\rho^{io} \left(\underline{u}_w - \underline{u}_\Sigma^o \right)]\!] \cdot \underline{n} = 0 \\[2mm] [\![\rho^{ig} \left(\underline{u}_g - \underline{u}_\Sigma^g \right)]\!] \cdot \underline{n} = 0 \end{array} \right\}, \quad i = 1, ..., N_C \tag{8.28}$$

which have to be fulfilled at shocks, Σ, where discontinuities of the intensive properties occur. Here, we have written

$$\underline{u}_\Sigma^\alpha \equiv \varepsilon S_\alpha \underline{v}_\Sigma, \alpha = w, o, g. \tag{8.29}$$

However, for simplicity, in what follows we will not discuss shocks.

Summing up the last two equations of Eq. (8.27) and making use of Eq. (8.26) we obtain

$$\left. \begin{array}{l} \dfrac{\partial \varepsilon S_w \rho^w}{\partial t} + \nabla \cdot \left(\rho^w \underline{u}_w \right) = g^w \\[3mm] \dfrac{\partial \varepsilon \left(S_o \rho^{io} + S_g \rho^{ig} \right)}{\partial t} + \nabla \cdot \left(\rho^{io} \underline{u}_o + \rho^{ig} \underline{u}_g \right) = g_E^{io} + g_E^{ig} \end{array} \right\}, \quad i = 1, ..., N_C. \tag{8.30}$$

Here, $g^i \equiv g_E^{io} + g_E^{ig}$ is the total volatile oil that goes into the reservoir through wells.

Standard versions of the compositional model incorporate the following constitutive equations:

1. Darcy's law for each phase, as given by Eq. (8.13);

2. The saturation identity of Eq. (8.15);

3. The distribution of each hydrocarbon component in each one of the oil and gas phases is subjected to the condition of *stable thermodynamic equilibrium*. This is expressed by means of the following N_C equalities:

$$f_{io}\left(p_o, \chi_{1o}, ..., \chi_{N_Co}\right) = f_{ig}\left(p_g, \chi_{1g}, ..., \chi_{N_Cg}\right), \quad i = 1, ..., N_C. \quad (8.31)$$

Here f_{io} and f_{ig} are the fugacity functions, which generally depend on the temperature, one of them for each component. The role of Eqs. (8.31) in *compositional models* is very important because they determine the interchange between phases of the mass of each component. The *stable thermodynamic equilibrium* condition of Eq. (8.31) can be derived by minimizing the *Gibbs free energy* of the compositional system (see [[1], [5], [4]]for more details); and

4. The capillary pressure relations, as explained in Section 7.4 and given by Eqs. (8.17) and (8.18).

The notation used in Eq. (8.31), as well as that of the developments that follow, requires clarification. The symbols $\xi_{i\alpha}$ and $\chi_{i\alpha}$ represent the *molar density* and the *mole fraction* of component i in phase α, respectively, while ξ_α is the *molar density* of phase α. They satisfy the following definitions :

$$\xi_{i\alpha} \equiv \frac{\rho^{i\alpha}}{W_i}, \quad \xi_\alpha \equiv \sum_{i=1}^{N_C} \xi_{i\alpha}, \text{ and } \chi_{i\alpha} \equiv \frac{\xi_{i\alpha}}{\xi_\alpha}. \quad (8.32)$$

Here, W_i is the *molecular weight* of component i. Therefore,

$$\rho^{i\alpha} = W_i \xi_{i\alpha} = W_i \xi_\alpha \chi_{i\alpha}. \quad (8.33)$$

Thereby, we observe that

$$\sum_{i=1}^{N_C} \chi_{io} = 1 \text{ and } \sum_{i=1}^{N_C} \chi_{ig} = 1. \quad (8.34)$$

The fact that the thermodynamic equilibrium conditions of Eq. (8.31) are expressed in terms of the mole fractions has a strong influence on the manner in which compositional models are usually formulated. Indeed, due to this fact, it is advantageous to express the balance equations themselves in terms of the mole fractions. Dividing each of Eqs. (8.30) by the molecular weight of the corresponding component, and making use of Eq. (8.33), one obtains:

$$\left. \begin{array}{l} \dfrac{\partial \varepsilon \xi_w S_w}{\partial t} + \nabla \cdot \left(\xi_w \underline{u}^w\right) = q_w \\[2ex] \dfrac{\partial \varepsilon \left(\chi_{io} \xi_o S_o + \chi_{ig} \xi_g S_g\right)}{\partial t} \\[2ex] + \nabla \cdot \left(\chi_{io} \xi_o \underline{u}^o + \chi_{ig} \xi_g \underline{u}^g\right) = \chi_{io} q_o + \chi_{ig} q_g \end{array} \right\}, \quad i = 1, ..., N_C. \quad (8.35)$$

Here,

$$q_w \equiv \frac{g^w}{W_w},$$

$$q_o \equiv \sum_{i=1}^{N_C} \frac{g_E^{io}}{W_i}, \quad \text{and}$$

$$q_g \equiv \sum_{i=1}^{N_C} \frac{g_E^{ig}}{W_i}. \qquad (8.36)$$

In summary, we have $N_C + 1$ balance equations given by Eq. (8.35) and the following constitutive laws:

1. Darcy's law of Eq. (8.13);

2. The saturation identity of Eq. (8.15);

3. The capillary forces constitutive law of Eq. (8.17);

4. The thermodynamic equilibrium conditions of Eq. (8.31); and

5. The mole fraction identities of Eq. (8.34).

Counting the number of equations derived from these constitutive laws, we have one from Eq. (8.15), two from Eq. (8.17), and another two from Eq. (8.34), while Eq. (8.31) gives N_C. As for Darcy's law of Eq. (8.13), it gives nine equations because we have three phases, and for each phase we have one vector equation, which is equivalent to three scalar equations. Thus, this gives a total of $N_C + 14$. Adding to this the $N_C + 1$ balance equations, it is seen that we have a system of $2N_C + 15$ equations. The manner of choosing the independent functions which are required to solve this system is not unique. However, a convenient choice of them is as follows: χ_{io}, χ_{ig}, $\underline{u}_\alpha$, p^α, and S_α, $\alpha = w, o, g$, and finally, $i = 1, ..., N_C$. When this system is subjected to appropriate boundary and initial conditions, a *complete model* capable of predicting the behavior of the reservoir system is obtained.

8.6 SUMMARY

In this chapter we considered the simulation of petroleum-reservoir behavior. After an introduction to the general concept of petroleum-reservoir exploitation, we presented a unified theory for EOR (enhanced oil recovery) modeling. The specific example of the well-know black oil model is then discussed in detail. The chapter ends with a presentation of the modeling strategy for compositional models which consider both species (components) and multiple phases.

EXERCISES

8.1   Derive Eq. (8.12) from Eq. (8.9).

8.2   The equations of the black-oil model as they appear in (8.9) are not amenable to numerical treatment. This is because it is not possible to evaluate the exchange terms g_g^{Go} and g_{Go}^g. In order to obtain a system of equations suitable for numerical treatment it is essential to eliminate such terms, as we have done when we derived Eq. (8.12) from Eq. (8.9). However, after the system of equations (8.12) has been solved, it is possible to estimate the exchange terms g_g^{Go} and g_{Go}^g. Explain how you would do that.

Note.   An adequate treatment of the problems that follow requires using the *jump condition* in the inhomogeneous form of Eq. (1.33):

$$[\![\phi S_w \rho_w (v_w - v_\Sigma)]\!] \cdot \underline{n} = 0 \text{ on } \Sigma \tag{8.37}$$

$$[\![\phi S_o \rho_{Oo} (v_o - v_\Sigma)]\!] \cdot \underline{n} = 0 \text{ on } \Sigma \tag{8.38}$$

$$[\![\phi S_o R_S \rho_{Oo} (v_o - v_\Sigma)]\!] \cdot \underline{n} = g_{\Sigma g}^{Go} \text{ on } \Sigma \tag{8.39}$$

$$[\![\phi S_g \rho_g (v_g - v_\Sigma)]\!] \cdot \underline{n} = g_{\Sigma Go}^g \text{ on } \Sigma. \tag{8.40}$$

Here, $g_{\Sigma g}^{Go}$ and $g_{\Sigma Go}^g$ represent the gas-mass exchange between the gas phase and the oil phase. Since the total gas mass is conserved, we have

$$g_{\Sigma g}^{Go} + g_{\Sigma Go}^g = 0. \tag{8.41}$$

Furthermore, the porosity is continuous and can be taken out of the square brackets.

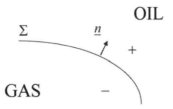

Figure 8.1   Gas front advancing into a region of unsaturated oil.

8.3   (**Motion of a gas front**). Assume that the water phase is absent. When a gas front advances into a region of unsaturated oil (see Fig. 8.1) a shock is formed at the gas front [[8]]. In such a case the gas-front does not advance into the region of unsaturated oil with the velocity of the gas; instead, it is retarded by a factor $0 < \eta \le 1$. More precisely,

$$(v_\Sigma - v_{o+}) \cdot \underline{n} = \eta (v_g - v_{o+}) \cdot \underline{n} \text{ on } \Sigma \tag{8.42}$$

where the *retardation factor* η is given by

$$\eta \equiv \left\{ 1 + [\![R_S]\!] \frac{S_{o+}\rho_{Oo+}}{S_g\rho_g} \right\}^{-1} \leq 1. \tag{8.43}$$

To prove this result the following hint can be used, but it is necessary to justify every step in it.

Hint: Equations (1.2) and (1.3) together imply that

$$\phi S_{o+}\rho_{Oo+} \left(\underline{v}_{o+} - \underline{v}_\Sigma \right) [\![R_S]\!] \cdot \underline{n} = g_{\Sigma g}^{Go} \text{ on } \Sigma. \tag{8.44}$$

The gas-front (see Fig. 8.1) is characterized by $(S_g)_- = 0$. Therefore,

$$\phi \left[\![S_g\rho_g \left(\underline{v}_g - \underline{v}_\Sigma \right)]\!] \right] \cdot \underline{n} = \phi S_g\rho_g \left(\underline{v}_g - \underline{v}_{o+} + \underline{v}_{o+} - \underline{v}_\Sigma \right) \cdot \underline{n} = g_{\Sigma Go}^g, \text{ on } \Sigma. \tag{8.45}$$

Furthermore, the fact that the porosity is continuous across the gas front Σ can be used.

Note. The black-oil model, in the form we have presented it in Section 8.4, is a very general model; that is, it contains as particular cases many models of interest. For example, when no gas is present in the system (i.e., when there is no gas phase, or dissolved gas), it reduces to an immiscible displacement model from which the Buckley-Leverett theory can be derived [[10], [2], [15], [3]]. This theory, in turn, has played a very important role in the evaluation of oil displacement by water (or some other displacing fluid) in reservoirs. The following four problems are devoted to deriving the essence of it. Capillary pressure and gravity effects are neglected throughout the discussions that follow.

8.4 Assuming that $g^w = g^{Oo} = 0$ in the first two equalities of Eq. (8.12), derive the equations

$$\frac{\partial}{\partial t} \left(\phi S_w \rho_w \right) + \nabla \cdot \left(\rho_w \underline{u}_w \right) = 0, \tag{8.46}$$

$$\frac{\partial}{\partial t} \left(\phi S_o \rho_{Oo} \right) + \nabla \cdot \left(\rho_{Oo} \underline{u}_o \right) = 0 \tag{8.47}$$

where the Darcy velocities are

$$\underline{u}_\alpha = \phi S_\alpha \underline{v}^\alpha; \ \alpha = w, o. \tag{8.48}$$

8.5 Define the *total Darcy velocity* by

$$\underline{u}_T = \underline{u}_w + \underline{u}_o. \tag{8.49}$$

Assuming that the porous medium and the fluids are incompressible (in particular that the fluid densities are independent of position and time) show that the total Darcy velocity fulfills

$$\nabla \cdot \underline{u}_T = 0. \tag{8.50}$$

Express the total Darcy velocity in terms of the pressure (when capillary pressure vanishes all phases have the same pressure) and obtain the partial differential equation satisfied by the pressure. What kinds of boundary conditions are required to determine the pressure?

8.6 This is a continuation of Exercise 8.5. Show that

$$\underline{u}_w = f_w \underline{u}_T \tag{8.51}$$

where

$$f_w(S_w) \equiv \frac{1}{1 + \dfrac{k_{ro}\mu_w}{k_{rw}\mu_o}}. \tag{8.52}$$

We have written $f_w(S_w)$ because the relative permeabilities generally are functions of the water saturation. Show that Eq. (1.10) reduces to

$$\frac{\partial S_w}{\partial t} + \phi^{-1} f_w' \underline{u}_T \cdot \nabla S_w = 0. \tag{8.53}$$

For this equation a *characteristic curve*, in space-time, is defined to be the trajectory of any particle whose velocity is $\phi^{-1} f_w' \underline{u}_T$. Prove that the water saturation remains constant along any characteristic curve.

8.7 Depending on the specific conditions of the reservoir system considered, the characteristic curves may or may not cross. When several characteristics, carrying different values of S_w, meet at a point, an incompatibility occurs and a continuous solution does not exist. Then a discontinuous solution has to be constructed. This is the mechanism of generation of a shock of the Buckley-Leverett type [[2],[8]]. Establish the jump conditions to be satisfied by such a shock. Furthermore, show that the velocity of the water front is given by

$$\underline{v}_\Sigma \cdot \underline{n} = \phi^{-1} \frac{[\![f_w]\!]}{[\![S_w]\!]} \underline{u}_T \cdot \underline{n}. \tag{8.54}$$

8.8 Show Eqs. (8.33) and (8.34).

8.9 Supply the details of the derivation of Eq. (8.9) from Eq. (8.1).

8.10 Supply the details of the derivation of Eq. (8.27) from Eq. (8.1).

REFERENCES

1. Bear, J., *Dynamics of Fluids in Porous Media*, Elsevier, New York, 1972.

2. Buckley, S.E. and M.C. Leverett, Mechanism of fluid displacement in sands, *Trans. AIME,* 146, 107-116, 1942.

3. Cardwell, W.T., The meaning of the triple value in noncapillary Buckley-Leverett theory, *Trans. AIME*, 216, 271-276, 1959.

4. Chen, Z., G. Huan, and Y. Ma, *Computational Methods for Multiphase Flows in Porous Media*, SIAM, Philadelphia, 2006.

5. Chen, Z., Formulations and numerical methods of the black oil model in porous media, *SIAM J. Numer. Anal.*, 38, (2), 489–514, 2000.

6. Herrera, I. and L. Chargoy, Shocks in solution gas-drive reservoirs, SPE Thirteenth Symposium in Reservoir Simulation, San Antonio, TX, Paper SPE 029136, 429-440, 1995.

7. Herrera, I. and G. Herrera, Unified formulation on enhanced oil-recovery method, *Geofís. Int.*, 50(1), 85-98, 2011.

8. Herrera, I., R. Camacho, and A. Galindo, Shock modelling in petroleum engineering, Chapter 7 of *Computational Methods for Moving Boundary Problems in Heat and Fluid Flow*, Eds., L.C. Wrobel and C.A. Brebbia, 143-170, CMP Books, San Francisco, CA, 1993.

9. Herrera, I., A. Galindo, and R. Camacho, Shock modelling in variable bubble point problems of petroleum engineering, *Computational Modelling of Free and Moving Boundary Problems*, Vol. 1: Fluid Flow, Eds. L.C. Wrobel and C.A. Brebbia, CMP Books, San Francisco, CA, 399-415, 1991.

10. Leverett, M.C., Capillary behaviour in porous solids, *Trans. AIME*, 142, 159-172,1941.

11. Peaceman, D.W., *Fundamentals of Numerical Reservoir Simulation,* Elsevier, New York, 1977.

12. Szilas, A.P., *Production and Transport of Oil and Gas*, second completely revised edition, Part A: Flow Mechanics and Production, Developments in Petroleum Science, Vol.18A, Elsevier, Amsterdam, 1985.

13. Szilas, A.P., *Production and Transport of Oil and Gas*, second completely revised edition, Part B: Gathering and Transport, Developments in Petroleum Science, Vol.18B, Elsevier, Amsterdam,1986.

14. Thomas, G.W., *Principles of Hydrocarbon Reservoir Simulation*, International Human Resources Development Corporation, Boston, 1982.

15. Welge, H. J., A simplified method for computing oil recovery by gas or water drive, *Petrol. Trans. AIME*, 195, 91-98, 1952.

CHAPTER 9

LINEAR ELASTICITY

9.1 INTRODUCTION

The grounds on which solid mechanics and the mechanics of (free) fluids are based share many aspects, as was explained in Chapter 2. As was seen there, the basic family of extensive properties for both mathematical models is the same: mass, linear momentum, angular momentum, kinetic energy and internal energy. When we assume from the outset, as we will do here, that the stress tensor is symmetric, the angular momentum can be deleted from this list.

A fundamental difference between solids and fluids lies in the kind of constitutive equations that characterize each one of these two broad classes of continuous systems. Both solids and fluids deform under the action of forces, and there is a correspondence between the forces that are applied and the deformation that is produced. The constitutive equations we are referring to specify such a correspondence and they are generally known as *stress-strain relations*. This chapter is devoted to some fundamental concepts of solid mechanics, while fluid mechanics is considered in the next.

There is a wide range of stress-strain relations that are applicable to solids. Two very general classes correspond to *elastic* and *viscoelastic* materials. Here, however,

Mathematical Modeling in Science and Engineering: An Axiomatic Approach.
By Ismael Herrera and George F. Pinder Copyright © 2012 John Wiley & Sons, Inc.

we only discuss elastic materials; furthermore, except for excursions in which we place the subject in a more general perspective, this chapter is devoted to the *linear theory of elasticity*.

Linear elasticity is one of the more successful theories of mathematical physics. Its pragmatic success in describing small deformations of many materials is uncontested. The origins of the three-dimensional theory go back to the beginning of the 19th century. Our presentation draws extensively from the very comprehensive work on the subject, by Gurtin [1972][12]]. However, frequently we have changed the point of view presented by Gurtin to make our presentation more in line with the needs of the audiences to which our book is intended.

9.2 ELASTIC SOLIDS

In what follows, the *material coordinates* $\underline{X}$ are identified with the initial positions and we recall that, for each $\underline{X}$, $p(\underline{X}, t)$ is the position vector of the particle $\underline{X}$ at time t. Then, we define the *displacement* to be

$$\underline{u}(\underline{X}, t) \equiv \underline{p}(\underline{X}, t) - \underline{X}. \tag{9.1}$$

We also write $\underline{\underline{F}} \equiv \nabla_{\underline{X}} \underline{p}$ to describe the *displacement gradient*; here, the notation is used for the gradient of the Lagrangian representation, taken with respect to the material coordinates. Later, the notation $\nabla_{\underline{x}}$ will be used for the gradient of the Euclidean representation, taken with respect to the position in the physical space. Eq. (9.1) is equivalent to

$$\underline{p}(\underline{X}, t) = \underline{X} + \underline{u}(\underline{X}, t). \tag{9.2}$$

The deformation gradient is defined by

$$\underline{\underline{H}}(\underline{X}, t) \equiv \nabla_{\underline{X}} \underline{u}(\underline{X}, t) = \underline{\underline{F}}(\underline{X}, t) - \underline{\underline{I}}. \tag{9.3}$$

Here, $\underline{\underline{I}}$ is the identity matrix.

The stress tensor can be visualized as shown in Fig. 9.1. The Cauchy stress tensor is second order and given by

$$\underline{\underline{\sigma}} = \begin{bmatrix} \sigma_{11} & \sigma_{12} & \sigma_{13} \\ \sigma_{21} & \sigma_{22} & \sigma_{23} \\ \sigma_{31} & \sigma_{32} & \sigma_{33} \end{bmatrix} = \begin{bmatrix} \sigma_{xx} & \sigma_{xy} & \sigma_{xz} \\ \sigma_{yx} & \sigma_{yy} & \sigma_{yz} \\ \sigma_{zx} & \sigma_{zy} & \sigma_{zz} \end{bmatrix}. \tag{9.4}$$

A component such as σ_{11} is the stress normal to the face located on the top of the cube and the component σ_{31} is the stress acting in the x_1 direction on the same face. A general class of materials, called *simple materials*, is defined by the condition that its *stress-strain relations* are such that the stress tensor at any given time, and at a given point, is fully determined by the displacement gradient at the same time, and the same point. Thus, for simple materials the stress tensor is independent of the

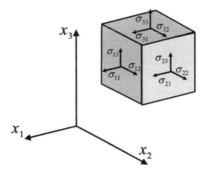

Figure 9.1 Diagrammatic representation of the stress tensor.

deformation rate. On the other hand for *viscoelastic materials*, also called *materials with memory*, the stress-tensor is a function not only of the rate of deformation, but rather may also depend on the whole past history of the deformation.

The mathematical expression of the above assumption, for *simple materials*, is

$$\underline{\underline{\sigma}} = \underline{\underline{S}}\left(\underline{F}\right) \tag{9.5}$$

where $\underline{\underline{S}}\left(\underline{F}\right)$ indicates the stress tensor for simple materials and is a function only of the displacement gradient $\underline{\underline{F}}$. *Elastic materials* are a subclass of simple materials, whose thorough study is identified with the term *finite elasticity*. Eq. (9.5) can be written using Eq. (9.3)as

$$\underline{\underline{\sigma}} = \underline{\underline{S}}\left(\underline{\underline{I}} + \underline{\underline{H}}\right). \tag{9.6}$$

If the solid has not been displaced from its *reference configuration*, $\underline{\underline{H}} = 0$ and Eq. (9.6) reduces to $\underline{\underline{\sigma}} = \underline{\underline{S}}\left(\underline{\underline{I}}\right)$. This stress matrix is called the *residual stress*. When the *reference configuration* is stress-free, $\underline{\underline{S}}\left(\underline{\underline{I}}\right) = \underline{\underline{0}}$.

9.3 THE LINEAR ELASTIC SOLID

For *linearly elastic solids* the reference configuration is always stress-free. Also, it is assumed that the displacement gradient is small. With this in mind, consider the following development

$$\frac{\partial u_i}{\partial X_j}\left(p\left(\underline{X}, t\right)\right) = \frac{\partial u_i}{\partial p_k} \frac{\partial p_k}{\partial X_j}. \tag{9.7}$$

Above, the summation convention is understood; that is, repeated indices are summed up over their range. This equation is usually written as

$$\frac{\partial u_i}{\partial X_j} = \frac{\partial u_i}{\partial x_k} \frac{\partial p_k}{\partial X_j} \tag{9.8}$$

or

$$\nabla_X \underline{u} = \nabla_x \underline{u} \nabla_X p \tag{9.9}$$

since $\underline{x} = \underline{p}(\underline{X}, t)$. But

$$\nabla_X \underline{u} = \underline{F} \tag{9.10}$$

so we have

$$\nabla_X \underline{u} = \nabla_x \underline{u} \underline{F}. \tag{9.11}$$

Eq. (9.3) implies that

$$\underline{F} = \underline{I} + \underline{H} \tag{9.12}$$

thus

$$\nabla_X \underline{u} = \nabla_x \underline{u} \left(\underline{I} + \underline{H} \right). \tag{9.13}$$

Rearranging and using a power series expansion for $\left((\underline{I} + \underline{H})^{-1} \right)$ yields

$$\nabla_x \underline{u} = \nabla_X \underline{u} \left(\underline{I} + \underline{H} \right)^{-1} = \underline{H} \left(\underline{I} - \underline{H} + O \left(\|\underline{H}\|^2 \right) \right) = \underline{H} + O \left(\|\underline{H}\|^2 \right) \tag{9.14}$$

Then Eqs. (9.14) and (9.3), together, imply that

$$\nabla_X \underline{u} = \nabla_x \underline{u} + O \left(\|\underline{H}\|^2 \right). \tag{9.15}$$

If we neglect terms of $O\left(\|\underline{H}\|^2 \right)$, we can replace $\nabla_X \underline{u}$ by $\nabla_x \underline{u}$ and hereinafter the notation $\nabla \underline{u}$ will be used to denote both $\nabla_X \underline{u}$ and $\nabla_x \underline{u}$.

Another notion from finite elasticity is the *finite strain tensor*, which considers deformations wherein both rotations and strains can be arbitrarily large; it is defined to be

$$\underline{\underline{D}}(\underline{X}, t) \equiv \frac{1}{2} \left(\underline{F}^T \underline{F} - \underline{I} \right)(\underline{X}, t). \tag{9.16}$$

In addition, one can define an *infinitesimal strain tensor*, which is basic to the linear theory of elasticity. The infinitesimal strain tensor, which describes displacements whose values and gradients are small when compared to unity, is defined to be

$$\underline{\underline{E}}(\underline{x}, t) \equiv \frac{1}{2} \left(\nabla \underline{u} + \nabla \underline{u}^T \right)(\underline{x}, t). \tag{9.17}$$

In the finite theory of elasticity, the importance of *rigid displacements* is that they do not produce strains. Correspondingly, in the linear theory of elasticity *infinitesimal rigid displacements*, as they will be here defined, do not produce stresses. A deformation $\underline{p}(\underline{x})$ is said to be a *finite rigid deformation* **at a point**, when the finite strain tensor vanishes there:

$$\underline{\underline{D}} = 0. \tag{9.18}$$

If a deformation is a finite rigid deformation **at every point of a body**, then:

$$\underline{p}(\underline{X}) = \underline{y}_0 + \underline{\underline{Q}} \left(\underline{X} - \underline{X}_0 \right). \tag{9.19}$$

Here, $\underline{X}_0$ and $\underline{y}_0$ are given vectors, while $\underline{\underline{Q}}$ is an orthogonal matrix (that is, a rotation matrix). When Eq.(9.19) holds, a direct computation using Eqs. (9.3) and (9.15) yields

$$\underline{\underline{F}} = \underline{\underline{Q}} \ and \ \nabla \underline{u} = \underline{\underline{Q}} - \underline{\underline{I}}. \tag{9.20}$$

In such a case, since by definition

$$\underline{\underline{Q}}\underline{\underline{Q}}^T = \underline{\underline{I}}$$

and, therefore, the finite strain tensor vanishes everywhere:

$$\underline{\underline{D}} = 0. \tag{9.21}$$

However, the infinitesimal strain tensor does not, that is:

$$\underline{\underline{E}} = -\frac{1}{2}\nabla \underline{u}^T \nabla \underline{u} = \frac{1}{2}\left(\underline{\underline{Q}} + \underline{\underline{Q}}^T\right) - \underline{\underline{I}}. \tag{9.22}$$

If the *displacement gradients* are small, $O\left(\varepsilon\right)$ say, then $\underline{\underline{E}} = O\left(\varepsilon^2\right)$. A displacement field $\underline{u}$ is said to be an infinitesimal rigid displacement at a point, when the infinitesimal strain tensor vanishes at that point. If a deformation is an *infinitesimal rigid deformation* at every point of a body, then:

$$\underline{p}\left(X\right) = \underline{y}_0 + \underline{\underline{W}}\left(\underline{X} - \underline{X}_0\right). \tag{9.23}$$

Here, $\underline{\underline{W}}$ is a skew symmetric matrix called the *tensor of infinitesimal rotation*. When discussing the linear theory of elasticity, we will make use of this equation with $\underline{X}$ replaced by $\underline{x}$ as was explained earlier.

A first assumption about the stress-strain relationship, which is special for linearly elastic solids, is that the stress tensor is a linear function of the displacement gradient. A second is that the stress tensor due to an infinitesimal rigid displacement vanishes. When the first of these conditions holds, the second one is equivalent to the condition that at any point and time the stress tensor is a linear function of the infinitesimal strain tensor. Furthermore, the assumption that the **stress is a linear function of the infinitesimal strain tensor** has the following mathematical expression:

$$\underline{\underline{\sigma}} = \underline{\underline{\underline{\underline{C}}}} : \nabla \underline{u} = \underline{\underline{\underline{\underline{C}}}} : \underline{\underline{E}}. \tag{9.24}$$

Here, $\underline{\underline{\underline{\underline{C}}}}$ is a fourth-order tensor, called the *elastic tensor*. It characterizes the linear function that determines the stress when the infinitesimal strain is given. We use the following notation for the elastic tensor:

$$\underline{\underline{\underline{\underline{C}}}} \equiv \left(C_{ijpq}\right). \tag{9.25}$$

Therefore, in indicial notation, and using the so-called summation convention in which repeated indexes are understood to be added up over their ranges (i, j, p, q range over $1, 2, 3$) Eq. (9.24) reads:

$$\sigma_{ij} = C_{ijpq}e_{pq}. \tag{9.26}$$

Here, and in what follows we use the notation

$$\underline{\underline{E}} = (e_{ij}).$$

As stated in the introduction section of this chapter, the angular-momentum balance is replaced by the condition that the stress tensor is symmetric. Therefore,

$$C_{ijpq} = C_{jipq}. \tag{9.27}$$

Another condition, which can be derived from thermodynamics considerations, is the condition that the elastic tensor be symmetric; that is,

$$\underline{\underline{C}} = \underline{\underline{C}}^T. \tag{9.28}$$

This latter assumption is tantamount to the condition that the line integral over any closed trajectory over the space of strains vanishes; that is,

$$\oint \underline{\underline{\sigma}} d\underline{\underline{E}} = 0. \tag{9.29}$$

In indicial notation Eq. (9.28) is:

$$C_{ijpq} = C_{pqij}. \tag{9.30}$$

Altogether, these assumptions can be summarized as follows: **The stress, at any point and time, is given by Eq. (9.24), where the elastic tensor satisfies the following symmetries**:

$$C_{ijpq} = C_{pqij} = C_{jipq} = C_{ijqp}. \tag{9.31}$$

9.4   MORE ON THE DISPLACEMENT FIELD DECOMPOSITION

In the linear theory of elasticity, the gradient of the displacements is given by

$$\underline{\underline{H}} = \begin{pmatrix} \frac{\partial u_1}{\partial x_1} & \frac{\partial u_1}{\partial x_2} & \frac{\partial u_1}{\partial x_3} \\ \frac{\partial u_2}{\partial x_1} & \frac{\partial u_2}{\partial x_2} & \frac{\partial u_2}{\partial x_3} \\ \frac{\partial u_3}{\partial x_1} & \frac{\partial u_3}{\partial x_2} & \frac{\partial u_3}{\partial x_3} \end{pmatrix}. \tag{9.32}$$

Which can be written as the sum of its symmetric and skew-symmetric part, that is:

$$\underline{\underline{H}} = \underline{\underline{E}} + \underline{\underline{W}}. \tag{9.33}$$

Here,

$$\underline{\underline{E}} = \frac{1}{2}\left(\nabla\underline{u} + \nabla\underline{u}^T\right) \; and \; \underline{\underline{W}} = \frac{1}{2}\left(\nabla\underline{u} - \nabla\underline{u}^T\right). \tag{9.34}$$

In indicial notation this is

$$e_{ij} = \frac{1}{2}\left(\frac{\partial u_i}{\partial x_j} + \frac{\partial u_j}{\partial x_i}\right) \; and \; w_{ij} = \frac{1}{2}\left(\frac{\partial u_i}{\partial x_j} - \frac{\partial u_j}{\partial x_i}\right). \tag{9.35}$$

The tensors $\underline{\underline{E}}$ and $\underline{\underline{W}}$ are, as noted earlier, the infinitesimal strain tensor and the tensor of infinitesimal rotation respectively (defined in Eq. (9.34)) .

The matrices whose indicial representations are

$$\frac{1}{2}\left(\delta_{ip}\delta_{jq} + \delta_{iq}\delta_{jp}\right) \; and \; \frac{1}{2}\left(\delta_{ip}\delta_{jq} - \delta_{iq}\delta_{jp}\right) \tag{9.36}$$

have the properties that

$$E_{ij} = \frac{1}{2}\left(\delta_{ip}\delta_{jq} + \delta_{iq}\delta_{jp}\right)\frac{\partial u_p}{\partial x_q} \; and \; W_{ij} = \frac{1}{2}\left(\delta_{ip}\delta_{jq} - \delta_{iq}\delta_{jp}\right)\frac{\partial u_p}{\partial x_q} \tag{9.37}$$

where δ_{ip} is the Kronecker delta. Another vector that is frequently referred to in the literature is the *vector of infinitesimal rotation*, defined by

$$\underline{\omega} \equiv \frac{1}{2}\nabla \times \underline{u}. \tag{9.38}$$

For any vector $\underline{a}$, the vector of infinitesimal rotation has the property that

$$\underline{\underline{W}}\,\underline{a} = \underline{\omega} \times \underline{a}. \tag{9.39}$$

9.5   STRAIN ANALYSIS

The *dilatation* is defined to be

$$\Theta \equiv \nabla \cdot \underline{u} = tr\underline{\underline{E}}. \tag{9.40}$$

It represents the change of volume per unit volume. Indeed, employing the divergence theorem we obtain

$$\int_{B(t)} \Theta dx = \int_{B(t)} \nabla \cdot \underline{u}dx = \int_{\partial B(t)} \underline{u} \cdot \underline{n}dx \tag{9.41}$$

A *homogeneous displacement field* is any that is independent of the position $\underline{x}$. Our interest in studying such infinitesimal displacement fields is because their analysis will contribute to increasing our insight into the distribution of displacement in the neighborhood of any point of a linearly elastic solid. We restrict our discussion to the case when the homogeneous displacement field is pure strain (that is, when the infinitesimal rigid displacement vanishes), since the infinitesimal rigid displacements were considered earlier.

Such an infinitesimal displacement field is given by

$$\underline{u}\,(\underline{x}) = \underline{\underline{E}}\,\underline{x}. \tag{9.42}$$

Now, by taking the coordinate axes in the proper directions, $\underline{\underline{E}}$ can be diagonalized, since it is a symmetric matrix. Then we can write

$$\underline{\underline{E}} = \begin{pmatrix} e_1 & 0 & 0 \\ 0 & e_2 & 0 \\ 0 & 0 & e_3 \end{pmatrix}. \tag{9.43}$$

And Eq. (9.42) reads:

$$\begin{cases} u_1 = e_1 x_1 \\ u_2 = e_2 x_2 \\ u_3 = e_3 x_3 \end{cases} . \tag{9.44}$$

We notice that the dilatation is given by

$$\Theta = e_1 + e_2 + e_3. \tag{9.45}$$

We now consider a few special cases.

Isochoric strain. A *pure-strain displacement field* is isochoric when

$$e_1 + e_2 + e_3 = 0. \tag{9.46}$$

Simple extension. This is the case when two of the proper values of $\underline{\underline{E}}$ are zero. Hence, for example, Eq.(9.44) becomes

$$\begin{cases} u_1 = e x_1 \\ u_2 = 0 \\ u_3 = 0 \end{cases} \tag{9.47}$$

In the light of this concept, the general *pure strain* case of Eq. (9.44) can now be interpreted as the superposition of three simple extensions in three orthogonal directions.

Simple shear. The simplest isochoric strain is *simple shear.* This is obtained by taking one of the proper values in Eq. (9.44) equal to zero and the other two of opposite sign. Thus, for example:

$$\begin{cases} u_1 = \chi x_1 \\ u_2 = -\chi x_2 \\ u_3 = 0 \end{cases} . \tag{9.48}$$

By a rotation of $\frac{\pi}{4}$ of the coordinate axes, Eq. (9.48) can be brought into a more familiar form; namely,

$$\begin{cases} u_1 = \chi x_2 \\ u_2 = \chi x_1 \\ u_3 = 0 \end{cases} . \tag{9.49}$$

Note, however, that the coordinates used in Eq. (9.49) are no longer proper for the matrix $\underline{\underline{E}}$.

Every isochoric strain is the superposition of two states of simple shear. Indeed, let $\underline{u}$ be given by Eq. (9.44) and assume $e_1 + e_2 + e_3 = 0$. Then, define

$$
\begin{cases}
u_1' = e_1 x_1 & u_1'' = 0 \\
u_2' = -e_1 x_2 \quad and \quad u_2'' = -e_3 x_2 \\
u_3' = 0 & u_3'' = e_3 x_3
\end{cases}
\tag{9.50}
$$

Then

$$
\underline{u} = \underline{u}' + \underline{u}''
\tag{9.51}
$$

because, $e_2 = -e_1 - e_3$.

Every state of pure strain can be represented, uniquely, as the superposition of a simple dilatation and an isochoric state of pure strain. Again, let $\underline{u}$ be given by Eq. (9.44) and define

$$
\begin{cases}
u_1' = \frac{1}{3}\Theta x_1 & u_1'' = e_1 x_1 - \frac{1}{3}\Theta x_1 \\
u_2' = \frac{1}{3}\Theta x_2 \quad and \quad u_2'' = e_2 x_2 - \frac{1}{3}\Theta x_2 \\
u_3' = \frac{1}{3}\Theta x_3 & u_3'' = e_3 x_3 - \frac{1}{3}\Theta x_3
\end{cases}
\tag{9.52}
$$

Then

$$
\underline{u} = \underline{u}' + \underline{u}''.
\tag{9.53}
$$

Every state of pure strain can be represented as the superposition of a simple dilatation and two states of simple shear. This is clear from the previous results.

9.6   STRESS ANALYSIS

As was seen in the introduction to this chapter, the balance of angular momentum is replaced by the condition that the stress tensor is symmetric. Thus, as in the previous section, in this one we analyze the forms that a symmetric matrix may take. However, the main difference between these forms is the physical interpretation of the results.

To simplify the physical interpretation, let us think of homogeneous stress fields, exclusively; that is, let us assume that the fields will be independent of position $\underline{x}$. The *traction* at a point of the boundary of a solid body is given by

$$
\underline{T}(\underline{x}) = \underline{\underline{\sigma}} \cdot \underline{n}.
\tag{9.54}
$$

Where $\underline{n}$ is the unit normal vector pointing outwards from the body.

By taking the coordinate axes in the proper directions, $\underline{\underline{\sigma}}$ can be diagonalized, since it is a symmetric matrix. Then

$$
\underline{\underline{\sigma}} = \begin{pmatrix}
\sigma_1 & 0 & 0 \\
0 & \sigma_2 & 0 \\
0 & 0 & \sigma_3
\end{pmatrix}.
\tag{9.55}
$$

When Eq. (9.55) holds, one has

$$\begin{cases} T_1 = \sigma_1 n_1 \\ T_2 = \sigma_2 n_2 \\ T_3 = \sigma_3 n_3 \end{cases} \tag{9.56}$$

Simple tension. This is the case when two of the proper values of $\underline{\underline{\sigma}}$ are zero, while the third one is greater than zero. Hence, for example, Eq. (9.44) becomes

$$\begin{cases} T_1 = \sigma n_1 \\ T_2 = 0 \qquad \sigma > 0. \\ T_3 = 0 \end{cases} \tag{9.57}$$

If $\sigma < 0$, the state of stress is one of *simple compression*.

Observe that when a body is in a state of simple tension the traction that is exerted on the body is:

$$\underline{T}(\underline{x}) = \underline{\underline{\sigma}} \cdot \underline{n} = \begin{pmatrix} \sigma & 0 & 0 \\ 0 & 0 & 0 \\ 0 & 0 & 0 \end{pmatrix} \begin{pmatrix} n_1 \\ 0 \\ 0 \end{pmatrix} = \begin{pmatrix} \sigma n_1 \\ 0 \\ 0 \end{pmatrix}. \tag{9.58}$$

Thus, for example, on the forward face of the cube illustrated in Fig. 9.2, the traction is

$$\underline{T} = \begin{pmatrix} \sigma \\ 0 \\ 0 \end{pmatrix}. \tag{9.59}$$

That is, it is perpendicular to that face and it is directed outwards (when $\sigma > 0$). As for the tractions on the faces perpendicular to this one in the same figure, they vanish.

In light of the concepts of simple tension and simple compression, the general stress state depicted by Eq. (9.56) can be interpreted as the superposition of three simple tensions or simple compressions, orthogonal to each other.

Isotropic states of stress. This is the case when all the proper values of the stress tensor are equal: $\sigma \equiv \sigma_1 = \sigma_2 = \sigma_3$. Therefore,

$$\underline{\underline{\sigma}} = \sigma \underline{\underline{I}}. \tag{9.60}$$

When $\sigma < 0$, the pressure is defined to be $p \equiv -\sigma$ and the stress state is said to be one of a *uniform pressure*. When, $\sigma > 0$ the stress state is one of *uniform isotropic tension*.

Pure shear. This is the case when one of the principal stresses vanishes and the other two are of opposite sign; that is,

$$\underline{\underline{\sigma}} = \begin{pmatrix} \tau & 0 & 0 \\ 0 & -\tau & 0 \\ 0 & 0 & 0 \end{pmatrix}. \tag{9.61}$$

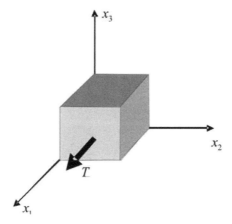

Figure 9.2   Tensile traction.

By a rotation of $\frac{\pi}{4}$, the stress matrix can be transformed into

$$\underline{\underline{\sigma}} = \begin{pmatrix} 0 & \tau & 0 \\ \tau & 0 & 0 \\ 0 & 0 & 0 \end{pmatrix}. \tag{9.62}$$

Every state of homogeneous stress is the superposition of an isotropic state of stress plus two states of pure shear. One can use the same argument as for the strain.

9.7   ISOTROPIC MATERIALS

When the material is isotropic, the number of independent coefficients reduces to only two and an elegant manner of deriving them is presented by Gurtin [12]. First, we show that

1. When an *isotropic strain* is applied, an *isotropic state of stress* is produced (that is, either a uniform pressure or isotropic tension);

2. The stress produced by an isochoric strain is traceless; and

When $\underline{\underline{E}}_0$ is an isochoric strain there exists a constant, called the *shear modulus*, such that

$$\underline{\underline{C}}\,\underline{\underline{E}} = 2\mu\underline{\underline{E}}. \tag{9.63}$$

By definition, the properties of isotropic materials are invariant under a change of coordinates. Thus in the following discussion it will be assumed that the system of coordinates is chosen in the principal directions of the applied strain, $\underline{\underline{E}}$. Furthermore, we observe that 3×3 matrices, which are symmetric, constitute an six dimensional linear space to be denoted by D. Furthermore, the operation $\underline{\underline{E}} : \underline{\underline{F}}$ defined for every $\underline{\underline{E}}$ and $\underline{\underline{F}}$ belonging to D, is given by

$$\underline{\underline{E}} : \underline{\underline{F}} = \sum_{i=1}^{3} \sum_{j=1}^{3} E_{ij} F_{ij}$$

and constitutes an inner product on D. On the other hand, the elastic tensor $\underline{\underline{C}}$ defines a linear transformation $\underline{\underline{C}} : D \to D$ of D into itself, since the stress tensor is also symmetric, that associates with each $\underline{\underline{E}} \in D$ the element $\underline{\underline{C}} : \underline{\underline{E}}$. This transformation is positive definite and self-adjoint with respect to the inner product just introduced.

When an isotropic strain is applied any direction is a *principal direction* of $\underline{\underline{E}}$, and therefore the stress produced $\underline{\underline{C}} : \underline{\underline{E}}$ is also invariant under a change of coordinates; in other words, an isotropic stress is produced (that is, either a *uniform pressure or isotropic tension*). This means, that

$$\underline{\underline{C}} : \underline{\underline{I}} = 3k\underline{\underline{I}}. \tag{9.64}$$

The coefficient k is called the *modulus of compression*. Clearly, k is a proper value and so $k > 0$.

Let us number the positions in the strain matrix as indicated next:

$$\begin{pmatrix} 1 & 4 & 6 \\ 0 & 2 & 5 \\ 0 & 0 & 3 \end{pmatrix}. \tag{9.65}$$

Then we can represent $\underline{\underline{C}}$ by

$$\underline{\underline{C}} \equiv (C_{ij}), i, j = 1, 2, ..., 6 \tag{9.66}$$

and $\underline{\underline{E}}$ by

$$\underline{\underline{E}} \equiv (E_1, ..., E_6). \tag{9.67}$$

With this notation the inner product introduced in Eq.9.63 is given in general by

$$\underline{\underline{E}} \cdot \underline{\underline{F}} = \sum_{i=1}^{3} E_i F_i + 2 \sum_{i=4}^{6} E_i F_i. \tag{9.68}$$

Then, the operation $\underline{\underline{C}} : \underline{\underline{E}}$ can be replaced by

$$\left(\underline{\underline{C}} \underline{\underline{E}}\right)_i = \sum_{j=1}^{3} C_{ij} E_j + 2 \sum_{j=4}^{6} C_{ij} E_j. \tag{9.69}$$

Let $\underline{\underline{E}}$ be a simple shear and choose the system of coordinates in the principal directions of $\underline{\underline{E}}$. Furthermore, assume that in such a system of coordinates

$$\underline{\underline{E}} = \begin{pmatrix} 1 & 0 & 0 \\ 0 & -1 & 0 \\ 0 & 0 & 0 \end{pmatrix}. \tag{9.70}$$

By making a change of coordinates such that x_1 and x_2 are interchanged, while x_3 is replaced by $-x_3$, and using the invariance under right-hand changes of coordinates, it is seen that $C_{41} = C_{42}$. Applying other cyclical replacements of $x_1 \to x_2 \to x_3 \to x_1$, one obtains

$$C_{11} - C_{12} = C_{22} - C_{21}. \tag{9.71}$$

When $\underline{\underline{E}}$ is given by Eq. (9.70), this shows that there exists a real number, denoted by 2μ, such that

$$\underline{\underline{\underline{C}}} : \underline{\underline{E}} = 2\mu \begin{pmatrix} 1 & 0 & 0 \\ 0 & -1 & 0 \\ 0 & 0 & 0 \end{pmatrix}. \tag{9.72}$$

Again μ is a proper value and so $\mu > 0$.

We observe that the only assumption made above was that $\underline{\underline{E}}$ is a simple shear. Therefore, any simple shear is a proper vector of the elasticity tensor and with the same proper value: 2μ. Furthermore, every isochoric strain can be expressed as the sum of two states of simple shear. Hence, the result just obtained implies that every isochoric strain is a proper vector of the elasticity tensor. Finally, every strain is the sum of an *isotropic state of strain* and an isochoric strain and we already know how to predict the stress produced by each one of these; thus, we are ready to develop explicit formulas for the stress-strain relations of an isotropic linearly-elastic material.

9.8 STRESS-STRAIN RELATIONS FOR ISOTROPIC MATERIALS

The *infinitesimal strain* is defined to be

$$\underline{\underline{E}} = \frac{1}{2} \left(\nabla \underline{u} + \nabla \underline{u}^T \right). \tag{9.73}$$

We write

$$\underline{\underline{E}} \equiv (E_{ij}) \ and \ E_{ij} = \frac{1}{2} \left(\delta_{ip}\delta_{jq} + \delta_{iq}\delta_{jp} \right) \frac{\partial u_i}{\partial x_j}. \tag{9.74}$$

A uniform dilatation with the same trace as $\underline{\underline{E}}$ is

$$\frac{1}{3} \delta_{ij} \delta_{pq} \frac{\partial u_p}{\partial x_q}. \tag{9.75}$$

Adding and subtracting this expression in Eq. (9.74), one obtains

$$E_{ij} = \frac{1}{2}\left(\delta_{ip}\delta_{jq} + \delta_{iq}\delta_{jp} - \frac{2}{3}\delta_{ij}\delta_{pq}\frac{\partial u_p}{\partial x_q}\right)\frac{\partial u_i}{\partial x_j} + \frac{1}{3}\delta_{ij}\delta_{pq}\frac{\partial u_p}{\partial x_q}. \tag{9.76}$$

Here, the expressions

$$\frac{1}{2}\left(\delta_{ip}\delta_{jq} + \delta_{iq}\delta_{jp} - \frac{2}{3}\delta_{ij}\delta_{pq}\frac{\partial u_p}{\partial x_q}\right)\frac{\partial u_i}{\partial x_j} \text{ and } \frac{1}{3}\delta_{ij}\delta_{pq}\frac{\partial u_p}{\partial x_q} \tag{9.77}$$

represent, respectively, an isochoric strain and an isotropic strain. Therefore,

$$
\begin{aligned}
\sigma_{ij} &= C_{ijpq}E_{pq} \\
&= C_{ijpq}\frac{\partial u_p}{\partial x_q} \\
&= \mu\left(\delta_{ip}\delta_{jq} + \delta_{iq}\delta_{jp} - \frac{2}{3}\delta_{ij}\delta_{pq}\right)\frac{\partial u_p}{\partial x_q} + k\delta_{ij}\delta_{pq}\frac{\partial u_p}{\partial x_q}. \tag{9.78}
\end{aligned}
$$

An algebraic manipulation of this equation yields

$$\sigma_{ij} = \lambda\frac{\partial u_p}{\partial x_q}\delta_{pq}\delta_{ij} + \mu\left(\delta_{ip}\delta_{jq} + \delta_{iq}\delta_{jp}\right)\frac{\partial u_p}{\partial x_q}. \tag{9.79}$$

The expression of Eq. (9.79) is due to Cauchy. When this strain-stress relation is used, λ and μ are referred to as the *Lamé constants*. Here,

$$\lambda \equiv k - \frac{2}{3}\mu \tag{9.80}$$

where μ is the *shear modulus*.

If a form in which the *strain* is explicitly used is desired, Eq. (9.79) can also be written as

$$\underline{\underline{\sigma}} = 2\mu\underline{\underline{E}} + \lambda\left(\text{tr } \underline{\underline{E}}\right)\underline{\underline{I}}. \tag{9.81}$$

Therefore,

$$C_{ijpq} = \lambda\delta_{pq}\delta_{ij} + \mu\left(\delta_{ip}\delta_{jq} + \delta_{iq}\delta_{jp}\right). \tag{9.82}$$

Alternative forms of expressing the stress-strain relations are the following:

$$\sigma_{ij} = \lambda E_{pp}\delta_{ij} + 2\mu E_{ij} = \lambda\Theta\delta_{ij} + 2\mu E_{ij} \tag{9.83}$$

$$\sigma_{ij} = \lambda\left(\frac{\partial u_p}{\partial x_p}\right)\delta_{ij} + 2\mu\left(\frac{\partial u_i}{\partial x_j} + \frac{\partial u_j}{\partial x_i}\right) \tag{9.84}$$

and

$$\underline{\underline{\sigma}} = \lambda\left(\nabla\cdot\underline{u}\right)\underline{\underline{I}} + \mu\left(\nabla\underline{u} + \nabla\underline{u}^T\right). \tag{9.85}$$

Other parameters that are also used to express the strain-stress relations of isotropic materials are the Young modulus E and Poisson ratio η (this is frequently denoted by σ, something we cannot do since that Greek letter has already been used for the stress tensor). The following equations state the relations between such parameters:

$$\lambda = \frac{\eta E}{(1+\eta)(1-2\eta)} \quad \mu = \frac{E}{2(1+\eta)}$$

$$\mu = \frac{E}{2(1+\eta)} \quad E = \frac{\mu(3\lambda+2\mu)}{\lambda+\mu} \qquad (9.86)$$

$$k = \lambda + \frac{2}{3}\mu \quad E = \frac{3\mu k}{\lambda+\mu}.$$

It should also be noticed that the solid is incompressible ($k = 0$) if, and only if, the Young modulus vanishes.

9.9 THE GOVERNING DIFFERENTIAL EQUATIONS

The balance equations presented in Sections 2.2 to 2.11, starting on page 23, are the starting points for the developments of this section. We assume that the body forces, $\underline{b}$, the heat flux, $\underline{q}$, and the heat sources, h, vanish. Then the balance equations take the form

$$\frac{\partial \rho}{\partial t} + \underline{v} \cdot \nabla \rho + \rho \nabla \cdot \underline{v} = 0$$

$$\rho \frac{\partial \underline{v}}{\partial t} + \rho \underline{v} \cdot \nabla \underline{v} - \nabla \cdot \underline{\underline{\sigma}} = \rho \underline{b} \qquad (9.87)$$

$$\rho \frac{\partial E}{\partial t} + \rho \underline{v} \cdot \nabla E = \underline{\underline{\sigma}} : \nabla \underline{v}.$$

Let ρ_0 and E_0 be two constants. Then a solution of this system of equations, when $\underline{b} = 0$, is

$$\rho = \rho_0$$

$$\underline{v} = 0 \qquad (9.88)$$

$$E_I = E_{I0}$$

where E_I is the internal energy.

Let ρ, $\underline{v}$, and E_I be such that $\rho - \rho_0$, $\underline{v}$, and $E_I - E_{I0}$, together with their gradients, $\nabla \underline{v}$, are small, $O(\varepsilon)$ say. Then it can be verified that the system of Eqs. (9.87) reduces to

$$\frac{\partial \rho}{\partial t} + \rho_0 \nabla \cdot \underline{v} = 0$$

$$\rho_0 \frac{\partial \underline{v}}{\partial t} - \nabla \cdot \underline{\underline{\sigma}} = \rho_0 \underline{b} \qquad (9.89)$$

$$\frac{\partial E}{\partial t} = 0$$

except for errors of order $O\left(\varepsilon^2\right)$. We observe that the assumptions above also require that $\underline{b}$ be $O\left(\varepsilon\right)$, since then

$$\rho\underline{b} = \rho_0\underline{b} + (\rho - \rho_0)\underline{b} = \rho_0\underline{b} + O\left(\varepsilon^2\right). \tag{9.90}$$

9.9.1 Elastodynamics

The balance equations used in linear elasticity are based on Eq. (9.89). Generally, one solves the balance of momentum equation first and the solution of the other two is then clear. So, in what follows, we will be dealing with a single equation that we shall write, since $v = \frac{\partial u}{\partial t}$, as

$$\rho\frac{\partial^2 \underline{u}}{\partial t^2} = \nabla \cdot \underline{\underline{\sigma}} + \rho\underline{b}. \tag{9.91}$$

Recalling Eq. (9.24) and applying matrix or indicial notation, Eq. (9.91) may be written as

$$\rho\frac{\partial^2 \underline{u}}{\partial t^2} - \nabla \cdot \left(\underline{\underline{C}} : \nabla\underline{u}\right) = \rho\underline{b} \tag{9.92}$$

or

$$\rho\frac{\partial^2 u_i}{\partial t^2} - \frac{\partial}{\partial x_j}\left(C_{ijpq}\frac{\partial u_p}{\partial x_q}\right) = \rho b_i. \tag{9.93}$$

The equations that are obtained for isotropic linearly elastic, homogeneous solids are known as the *Navier equations*. Taking Eq. (9.82) into consideration, they are

$$\rho\frac{\partial^2 \underline{u}}{\partial t^2} - (\lambda + \mu)\,\nabla\nabla \cdot \underline{u} - \mu\Delta\underline{u} = \rho\underline{b} \tag{9.94}$$

or

$$\rho\frac{\partial^2 u_i}{\partial t^2} - (\lambda + \mu)\,\frac{\partial}{\partial x_i}\nabla \cdot \underline{u} - \mu\Delta u_i = \rho b_i \tag{9.95}$$

where $\Delta\underline{u}$ is the Laplacian operator.

9.9.2 Elastostatics

When the elastic system is in equilibrium, the displacements are time independent and Eq. (9.92) reduces to

$$\nabla \cdot \left(\underline{\underline{C}} : \nabla\underline{u}\right) = -\rho\underline{b} \tag{9.96}$$

or

$$\frac{\partial}{\partial x_j}\left(C_{ijpq}\frac{\partial u_p}{\partial x_q}\right) = -\rho b_i. \tag{9.97}$$

For isotropic linearly elastic solids, which are homogeneous, Eq. (9.94) becomes

$$(\lambda + \mu)\,\nabla\nabla \cdot \underline{u} + \mu\Delta\underline{u} = -\rho\underline{b} \tag{9.98}$$

or

$$(\lambda + \mu) \frac{\partial}{\partial x_i} \nabla \cdot \underline{u} + \mu \Delta u_i = -\rho b_i. \tag{9.99}$$

By integration of the first expression of Eq. (9.89), an evaluation of the change of density due to the deformation of the solid is obtained:

$$\frac{\rho - \rho_0}{\rho_0} = -\nabla \cdot \underline{u}. \tag{9.100}$$

9.10 WELL-POSED PROBLEMS

We recall that well-posed problems are those which assure the existence and uniqueness of a solution. Furthermore, such a solution must depend continuously on the data of the problem.

9.10.1 Elastostatics

In this sub section we consider a domain Ω and its boundary $\partial \Omega$. Furthermore, the boundary is decomposed into two disjoint parts, $\partial_1 \Omega$ and $\partial_2 \Omega$, such that $\partial \Omega = \partial_1 \Omega \cup \partial_2 \Omega$. It is assumed that two vector-valued functions, $\underline{u}_{\partial 1}$ and $\underline{T}_{\partial 2}$, are defined on $\partial_1 \Omega$ and on $\partial_2 \Omega$, respectively. The *mixed problem of elastostatics* is to find a displacement field $\underline{u}$, defined in Ω, such that it fulfills Eq. (9.96) together with the boundary conditions

$$\underline{u} = \underline{u}_{\partial 1} \text{ on } \partial_1 \Omega$$
$$\tag{9.101}$$
$$\underline{\sigma} \cdot \underline{n} = \underline{T}_{\partial 2} \text{ on } \partial_2 \Omega.$$

The first of Eqs. (9.101) is called the *displacement conditions*, while the second one is known as the *traction conditions*. In Eq. (9.101) it is understood that

$$\underline{\underline{\sigma}} = \underline{\underline{\sigma}}(\underline{u}) = \underline{\underline{C}} : \nabla \underline{u}. \tag{9.102}$$

When $\partial_1 \Omega$ is non-void (actually, has non-zero area), this problem is well-posed (for a more careful formulation of this problem, see reference [12]). When $\partial_1 \Omega$ is void, there is more than one solution of the problem. Indeed, given a solution, another one is obtained by adding to it any infinitesimal rigid body motion (that is, the sum of a translation plus an infinitesimal rotation).

9.10.2 Elastodynamics

The problem to be considered in this sub section is defined in a space domain Ω and a finite (or infinite) time interval. In addition to the notation introduced before, we assume that $\underline{u}_0$ and $\underline{v}_0$ are two vector-valued functions defined in Ω.

The *mixed (initial-boundary-value) problem of elastodynamics* is to find a displacement field $\underline{u}$, defined in Ω, such that it fulfills Eq. (9.92) together with the *mixed boundary conditions*

$$\left.\begin{array}{c} \underline{u} = \underline{u}_{\partial 1} \text{ on } \partial_1\Omega \\ \underline{\underline{\sigma}} \cdot \underline{n} = \underline{T}_{\partial 2} \text{ on } \partial_2\Omega \end{array}\right\} \text{ for } t > 0 \tag{9.103}$$

and the *initial conditions*

$$\left.\begin{array}{c} \underline{u} = \underline{u}_0 \\ \dfrac{\partial \underline{u}}{\partial t} = \underline{v}_0 \end{array}\right\}, \ in \ \Omega \ for \ t = 0. \tag{9.104}$$

As here formulated, this problem is posed in an infinite time interval; however, as said before, this interval can be replaced by a finite interval in the formulation of the problem.

There are other problems of interest; albeit they will not be considered here, that should be mentioned. One is the *contact problem* and another is the *purely initial value problem*. The latter problem is well-posed in a domain that shrinks as time goes by; it shrinks with a velocity that equals the maximum velocity of elastic waves.

9.11 REPRESENTATION OF SOLUTIONS FOR ISOTROPIC ELASTIC SOLIDS

Every vector field can be represented in the form

$$\underline{u}\,(\underline{x}, t) = \underline{v}\,(\underline{x}, t) + \nabla\phi\,(\underline{x}, t)\,. \tag{9.105}$$

Here ϕ is a scalar function, while $\underline{v}$ is an \incompressible" (or isochoric) vector field that is such that

$$\nabla \cdot \underline{v} = 0. \tag{9.106}$$

So when we adopt this representation for $\underline{u}$ and substitute it in Eq. (9.94), with $\underline{b} = 0$, it gives rise to the following pair of equations:

$$\frac{1}{\alpha^2}\frac{\partial^2\phi}{\partial t^2} = \Delta\phi$$

$$\frac{1}{\beta^2}\frac{\partial^2\underline{v}}{\partial t^2} = \Delta\underline{v}. \tag{9.107}$$

Here it is understood that $\underline{v}$ fulfills Eq. (9.106), while

$$\alpha \equiv \left(\frac{\lambda + 2\mu}{\rho}\right)^{\frac{1}{2}} \text{ and } \beta \equiv \left(\frac{\mu}{\rho}\right)^{\frac{1}{2}}. \tag{9.108}$$

The latter parameters are known as the velocities of P (volumetric) and Q (shear) waves, respectively.

In view of the fact that $\underline{v}$ fulfills Eq. (9.106), one can write

$$\underline{v} = \nabla \times \underline{\psi}. \qquad (9.109)$$

Here, $\underline{\psi}$ is a vector field usually called the *vectorial potential* of $\underline{v}$. Replacing $\underline{v}$ by $\nabla \times \underline{\psi}$ in Eq. (9.105), the representation becomes

$$\underline{u}\,(\underline{x}, t) = \nabla \phi\,(\underline{x}, t) + \nabla \times \underline{\psi}. \qquad (9.110)$$

We observe that when Eq. (9.110) holds, the following relations are satisfied:

$$\nabla \cdot \underline{u}\,(\underline{x}, t) = \Delta \phi\,(\underline{x}, t) \text{ and } \nabla \times \underline{u}\,(\underline{x}, t) = \nabla \times \nabla \times \underline{\psi}\,(\underline{x}, t). \qquad (9.111)$$

Making use of Eq. (9.109), the second of Eqs. (9.107) becomes

$$\frac{1}{\beta^2} \frac{\partial^2\,(\nabla \times \underline{\psi})}{\partial t^2} = \Delta\,(\nabla \times \underline{\psi}) \qquad (9.112)$$

or

$$\nabla \times \left(\frac{1}{\beta^2} \frac{\partial^2 \underline{\psi}}{\partial t^2} - \Delta \underline{\psi} \right) = 0. \qquad (9.113)$$

In particular, if $\underline{\psi}$ satisfies

$$\frac{1}{\beta^2} \frac{\partial^2 \underline{\psi}}{\partial t^2} - \Delta \underline{\psi} = 0 \qquad (9.114)$$

then, the second of Eqs. (9.107) is satisfied.

Summarizing, if

$$\frac{1}{\alpha^2} \frac{\partial^2 \phi}{\partial t^2} = \Delta \phi$$
$$\qquad (9.115)$$
$$\frac{1}{\beta^2} \frac{\partial^2 \underline{\psi}}{\partial t^2} - \Delta \underline{\psi} = 0.$$

then the representation of Eq. (9.110) yields a displacement field that fulfills the homogeneous equation of elastodynamics, Eq. (9.94). In studies of elastic wave propagation, in particular in seismology [9], this representation of the displacement field in terms of both a scalar and a vector potential is widely used.

9.12 SUMMARY

In this chapter we consider linear elastic solids. After introducing a few preliminary definitions we introduce the concept of the linear elastic solid. The ensuing discussion considers the displacement field decomposition followed by the analysis of stress and of strain. Focusing on isotropic materials the stress-strain relationships are introduced. Thereafter the governing differential equations are presented, leading into a discussion of elastodynamics and elastostatics. We then define the conditions needed for a well-posed problem and in the final section present a representation of solutions for isotropic elastic solids.

EXERCISES

9.1   Let an elastic body be at rest and the positions of its particles be given by

$$p(\underline{X}) = \underline{a} + \underline{\underline{P}}\underline{X}. \tag{9.116}$$

Here $\underline{a} = (a_1, a_2, a_3)$, while

$$\underline{\underline{P}} \equiv (P_{ij}). \tag{9.117}$$

a) Write Eq. (9.116) in indicial notation:
b) Compute the displacement $\underline{u}$ at any *material particle* $\underline{X}$:
c) Evaluate the displacement of a material-particle located at a point $\underline{x}$ of the physical space:
d) Under what conditions does Eq. (9.116) represent a translation?
e) Under what conditions is it a rotation?

9.2   In linear algebra, a matrix is said to be *unitary* when

$$\underline{\underline{U}}^T\underline{\underline{U}} = \underline{\underline{I}}. \tag{9.118}$$

Here $\underline{\underline{I}}$ is the identity matrix. Write this equation in indicial notation and interpret this equation as the ortho normality conditions for the row vectors of $\underline{\underline{U}}^T\underline{\underline{U}} = \underline{\underline{I}}$.

9.3   Explain under what conditions Eq. (9.116) represents a rotation. What condition is satisfied by the determinant of the matrix $\underline{\underline{P}}$?

9.4   Evaluate the *displacement gradient,*[1] the *deformation gradient,* and the gradient with respect to the position in the physical space ($\nabla_{\underline{x}}\underline{u}$) when the positions of the particles are given by Eq. (9.117).

9.5   As explained in Section 9.2, a *simple material* is any for which the stress tensor, at any time, is exclusively a function of the displacement gradient, at that given time. The mathematical expression of this assumption is

$$\underline{\underline{\sigma}} = \underline{\underline{S}}\left(\underline{\underline{F}}\right). \tag{9.119}$$

One aspect of the usefulness (and beauty) of expressing an assumption in mathematical terms is its *wholeness*; indeed, the conditions that a solid material need to satisfy to be classified as a *simple material* are fully contained in Eq. (9.119). However, in order to master any physical subject sustained in a mathematical theory, it is necessary to fully understand the physical implications of the mathematical assumptions on which it is sustained. Although the mathematical expression for the basic assumption we have adopted for simple materials is very simple in its form, in this exercise we

---

[1]In Section 9.2 we have presented the standard definition (standard in the sense that it corresponds to the nomenclature that is usual in elasticity treatises [12]). We point out, however, that such a definition may be confusing in some instances, because to obtain it the gradient with respect to the *material coordinates* is applied to the position function: $p(\underline{X}, t)$.

invite the reader to explore the physical implications of Eq. (9.119). For example, assume that a simple material has been subjected to deformations in the past and say how such events in the history of the material would affect the stress tensor observed today.

9.6 Due to the *polar decomposition theorem,* the *displacement gradient* can be written as

$$\underline{\underline{F}} = \underline{\underline{Q}}\underline{\underline{V}} \tag{9.120}$$

where $\underline{\underline{Q}}$ is a rotation, and $\underline{\underline{V}}$ is a symmetric matrix. Obtain $\underline{\underline{Q}}$ and $\underline{\underline{V}}$ when the *displacement gradient* is

$$\underline{\underline{F}} = \begin{pmatrix} \sqrt{3} & \sqrt{3} & \sqrt{3} \\ \sqrt{2} - \frac{1}{\sqrt{2}} & -\frac{1}{\sqrt{2}} \\ 0 & 1 & -1 \end{pmatrix} \tag{9.121}$$

The *proper values* and *proper vectors* of $\underline{\underline{V}}$ are said to be the *principal values of strain* and *principal strain vectors,* respectively. Obtain them for the case when $\underline{\underline{F}}$ is given by Eq. (9.121).

9.7 Use the polar decomposition theorem to show that the *finite strain tensor* vanishes if and only if the displacement gradient is a rotation.

9.8 Show that the condition $C_{ijpq} = C_{pqij}$ holds if and only if

$$\oint C_{ijpq} de_{pq} = 0 \tag{9.122}$$

for every closed path in the e_{pq} *space.*

9.9 Assume that the state is one of *pure strain* and let the *infinitesimal strain* be

$$\underline{\underline{E}} = \begin{pmatrix} 5 & -1 & -1 \\ -1 & 3 & 1 \\ -1 & 1 & 3 \end{pmatrix} . \tag{9.123}$$

Express $\underline{\underline{E}}$ as the superposition of three simple extensions. In particular, show that $e_1 = 2$, $e_2 = 3$, and $e_3 = 6$. Represent this state of pure strain as the superposition of a simple dilatation and two states of simple shear.

9.10 Assume that the state is one of pure strain and let the infinitesimal strain be

$$\underline{\underline{E}} = \frac{1}{6} \begin{pmatrix} 2e_2 + 4e_3 & 2(e_2 - e_3) & 2(e_2 - e_3) \\ 2(e_2 - e_3) & 3e_1 + 2e_2 + e_3 & -3e_1 + 2e_2 + e_3 \\ 2(e_2 - e_3) & -3e_1 + 2e_2 + e_3 & 3e_1 + 2e_2 + e_3 \end{pmatrix} . \tag{9.124}$$

Show that the principal values of strain are e_1, e_2, and e_3. Represent this state of pure strain as the superposition of a simple dilatation and two states of simple shear.

9.11 Show Eq. (9.37).

9.12 Prove that Eqs. (9.63) and (9.64) imply, respectively, that 2μ is a proper value of the elastic tensor of multiplicity 5, while $3k$ has multiplicity 1. Use this result to show that any state of pure strain that is proper for the elastic tensor is either an isotropic state of strain or is an isochoric state of strain.

REFERENCES

1. Atanackovic, T.M. and A. Guran, *Theory of Elasticity for Scientists and Engineers,* Springer-Verlag, Berlin, 2000.

2. Atkin, R.J., *An Introduction to the Theory of Elasticity*, Longman, London, 1980.

3. Ciarlet, P.G., *Mathematical Elasticity*, Vol. II: Theory of Plates, Amsterdam, 1997.

4. Ciarlet, P.G., *Élasticité Tridimensionnelle*, Collection RMA1, 1986.

5. Coleman, B.D., M.E. Gurtin, I. Herrera, and C. Truesdell, *Wave Propagation in Dissipative Materials,* Springer-Verlag, New York, 1965.

6. Emanuel, G., *Analytical Fluid Dynamics*, CRC Press, Boca Raton, FL, 2001.

7. Eringen, C. and S. Suhubi, *Elastodynamics,* Academic Press, New York, 1975.

8. Evans, L.C., *Partial Differential Equations*, Graduate Studies in Mathematics, Vol. 19, American Mathematical Society, Providence, RI, 1998.

9. Ewing, W.M., W.S. Jardewsky, and F. Press, *Elastic Waves in Layered Media*, McGraw-Hill, New York, 1957.

10. Fanchi, J.R., *Shared Earth Modeling*, Butterworth Heinemann and Elsevier Science, New York, 2002.

11. Grove, D.B. and K.G. Stollenwerk, *Computer Model of One-Dimensional Equilibrium-Controlled Sorption Processes,* U.S. Geological Survey Water-Resources Investigations Report 84-4059, 1984.

12. Gurtin, M.E., The linear theory of elasticity, *Handbuch der Physik*, Vol. VIa/2, Ed. S. Flugge, Springer-Verlag, Berlin, 1972.

13. Gurtin, M. E., *An Introduction to Continuum Mechanics*, Academic Press, New York, 1981.

14. Herrera, I. and M.E. Gurtin, A correspondence principle for viscoelastic wave propagation, *Quart. of Appl. Math*, 22(4), 361-364, 1965.

15. Keller, H.B., Propagation of stress discontinuities in inhomogeneous elastic media, *SIAM Rev.*, 6, 356-382, 1964.

16. Landau, L.D. and F.M. Lifschitz, *Theory of Elasticity*, Pergamon Press, London, 1959.

17. Malvern, L.E., *Introduction to the Mechanics of a Continuous Medium*, Facsimile edition, Prentice- Hall, Englewood Cliffs, NJ, 1977.

18. Marsden, J.E. and T.R.J. Hughes, *Mathematical Foundations of Elasticity*, reprint edition, Dover, New York, 1994.

19. Muskhelishvili, N.L., *Some Basic Problems of the Mathematical Theory of Elasticity*, 3rd rev. and augmented ed., trans. from the Russian by J.R.M. Radok, Noordhoff, Groningen, The Netherlands, 1953.

20. Sokolnikoff, I.S., *Mathematical Theory of Elasticity,* McGraw-Hill, New York, 1956.

21. Yeh, H. and J.L. Abrams, *Principles of Mechanics of Solids and Fluids*, Vol. 1: Particle and Rigid-Body Mechanics, McGraw-Hill, New York, 1960.

CHAPTER 10

FLUID MECHANICS

10.1 INTRODUCTION

As we have seen in Chapter 2, the models of the mechanics of solids and fluids
derive from the same family of extensive properties. The difference between them
stems from their stress-strain relations. For solids, as we have seen in Chapter 9 for
simple materials, the stress tensor at a point is fully determined by the deformation
existing at that same point at the same time, and it is independent of the rate at which
deformation occurs. On the other hand, the stresses in a fluid depend strongly on the
rate of deformation.

The starting point for the discussions presented in this chapter is the set of balance
equations describing classical continuous systems first presented in Chapter 2. We
now list them in the forms more commonly used.

1. **Equation of continuity**

$$\frac{\partial \rho}{\partial t} + \nabla \cdot (\rho \underline{v}) = 0 \tag{10.1}$$

or

$$\frac{D\rho}{Dt} + \rho \nabla \cdot \underline{v} = 0. \tag{10.2}$$

Mathematical Modeling in Science and Engineering: An Axiomatic Approach.
By Ismael Herrera and George F. Pinder Copyright © 2012 John Wiley & Sons, Inc.

2. **Balance of linear momentum**

$$\rho \frac{\partial \underline{v}}{\partial t} + \rho \underline{v} \cdot \nabla \underline{v} - \nabla \cdot \underline{\underline{\sigma}} - \rho \underline{b} = 0 \qquad (10.3)$$

or

$$\rho \frac{D \underline{v}}{Dt} - \nabla \cdot \underline{\underline{\sigma}} - \rho \underline{b} = 0. \qquad (10.4)$$

3. **Balance of angular momentum**

$$\underline{\underline{\sigma}} = \underline{\underline{\sigma}}^T \qquad (10.5)$$

or

$$\sigma_{ij} = \sigma_{ji}. \qquad (10.6)$$

and

4. **Balance of energy**

$$\rho \frac{DE}{Dt} = \nabla \cdot \underline{q} + \rho h + \underline{\underline{\sigma}} : \nabla \underline{v}. \qquad (10.7)$$

10.2 NEWTONIAN FLUIDS: STOKES' CONSTITUTIVE EQUATIONS

This chapter is devoted to the study of Newtonian fluids, exclusively. The stress-deformation relations of such fluids fulfill the Stokes' axioms [6] which we discuss below.

To start with, we recall that the tractions, that is, the forces per unit area that act on the surface of a body, are given by

$$\underline{T} = \underline{\underline{\sigma}} \cdot \underline{n}. \qquad (10.8)$$

The normal component of $\underline{T}$ is the normal stress, while the tangential component is the shear stress. The first of *Stokes' axioms* states that when the fluid is at rest (so that $\nabla \cdot \underline{v} = 0$), no *shear stresses* are produced. Therefore, at rest, the tractions are necessarily orthogonal to the body surface. Furthermore, they are directed toward the body interior; that is, their sense is opposed to that of the unit normal vector. The mathematical expression of these conditions is

$$\underline{T} = -p\underline{n}. \qquad (10.9)$$

Here $p \geq 0$ and it is called the *hydrostatic pressure*. We observe that, in particular, the first Stokes' axiom implies that Newtonian fluids cannot withstand tensile stresses. Comparing Eqs. (10.8) and (10.9), it is seen that at rest

$$\underline{\underline{\sigma}} = -p\underline{\underline{I}}. \qquad (10.10)$$

In indicial notation this equation is

$$\sigma_{ij} = -p\delta_{ij}. \qquad (10.11)$$

The tensor of *viscous stress* is obtained when we subtract the stress tensor due to hydrostatic pressure from the stress tensor; that is, we define the viscous stress to be

$$\underline{\underline{\tau}} \equiv \underline{\underline{\sigma}} - \left(-p\underline{\underline{I}}\right) = \underline{\underline{\sigma}} + p\underline{\underline{I}}. \tag{10.12}$$

Therefore, we write

$$\underline{\underline{\sigma}} = \underline{\underline{\tau}} - p\underline{\underline{I}}. \tag{10.13}$$

A further *Stokes' axiom* states that *the viscous stress is a linear function of the gradient of fluid-particle velocities,* $\nabla \underline{v}$. In order to derive an explicit expression for the constitutive relations of *Newtonian fluids* from this assumption, we will carry out an analysis of the gradient of fluid-particle velocities, which is similar to the analysis of the deformation gradient that was done in Chapter 9.

First, we decompose it into its symmetric $\left(\underline{\underline{\varepsilon}}\right)$ and skew-symmetric $\left(\underline{\underline{\omega}}\right)$ parts:

$$\nabla \underline{v} = \underline{\underline{\varepsilon}} + \underline{\underline{\omega}}. \tag{10.14}$$

Here

$$\underline{\underline{\varepsilon}} \equiv \frac{1}{2}\left(\nabla\underline{v} + \nabla\underline{v}^T\right) \text{ and } \underline{\underline{\omega}} \equiv \frac{1}{2}\left(\nabla\underline{v} - \nabla\underline{v}^T\right). \tag{10.15}$$

The tensors $\underline{\underline{\varepsilon}}$ and $\underline{\underline{\omega}}$ are called the *strain rate* and the *rotation rate*, respectively. It is assumed that the stresses are only produced by the strain rate, while the rotation rate does not produce stresses. Hence,

$$\tau_{ij} = c_{ijpq}\frac{\partial v_p}{\partial x_q} = c_{ijpq}\varepsilon_{pq} \tag{10.16}$$

since, as noted, the rotation rate does not produce stresses.

For isotropic fluids, one arrives at the following expression for the coefficients occurring in Eq. (10.16):

$$c_{ijpq} = \lambda\delta_{ij}\delta_{pq} + \mu\left(\delta_{ip}\delta_{jq} + \delta_{iq}\delta_{jp}\right) \tag{10.17}$$

where λ and μ are invariant scalars, which depend on the thermodynamic state variables, such as temperature, and are known as the first and second viscosity coefficients, respectively. Notice the similarity with the stress-strain relations for linearly elastic materials that are isotropic, given in Eq. (9.81), albeit conceptually the parameters λ and μ, as here defined for fluids differ drastically from the Lamé constants for solids defined in Chapter 9.

Substituting Eq. (10.17) into Eq. (10.16) and carrying out the operations, one obtains

$$\tau_{ij} = \lambda\frac{\partial v_p}{\partial x_p}\delta_{ij} + \mu\left(\frac{\partial v_i}{\partial x_j} + \frac{\partial v_j}{\partial x_i}\right) \tag{10.18}$$

$$= \lambda\frac{\partial v_p}{\partial x_p}\delta_{ij} + 2\mu\varepsilon_{ij}. \tag{10.19}$$

Using matrix notation, this is

$$\underline{\underline{\tau}} = \lambda \frac{\partial v_p}{\partial x_p} \underline{\underline{I}} + 2\mu \underline{\underline{\varepsilon}} \tag{10.20}$$

or

$$\underline{\underline{\tau}} = \lambda \left(\nabla \cdot \underline{v} \right) \underline{\underline{I}} + 2\mu \underline{\underline{\varepsilon}}. \tag{10.21}$$

We now use Eq. (10.13) to incorporate the hydrostatic pressure and obtain

$$\underline{\underline{\sigma}} = \left(\lambda \nabla \cdot \underline{v} - p \right) \underline{\underline{I}} + 2\mu \underline{\underline{\varepsilon}} \tag{10.22}$$

or

$$\underline{\underline{\sigma}} = \left(\lambda \nabla \cdot \underline{v} - p \right) \underline{\underline{I}} + \mu \left(\nabla \underline{v} + \nabla \underline{v}^T \right). \tag{10.23}$$

Going back to using indicial notation, we can write Eq. (10.23) as

$$\sigma_{ij} = \left(\lambda \frac{\partial v_p}{\partial x_p} - p \right) \delta_{ij} + \mu \left(\frac{\partial v_i}{\partial x_j} + \frac{\partial v_j}{\partial x_i} \right). \tag{10.24}$$

Equations (10.22) to (10.24) are expressions of the *Stokes' constitutive equations* for an isotropic fluid with *Newtonian viscosity*. They are fully determined when the first and second viscosity coefficients, λ and μ, are specified.

10.3 NAVIER-STOKES EQUATIONS

In fluid mechanics the equations of balance of linear momentum for a Newtonian fluid are known as the *Navier-Stokes equations*. To derive them, we first notice that in view of Eq. (10.24), one has

$$\frac{\partial \sigma_{ij}}{\partial x_j} = \frac{\partial}{\partial x_i} \left\{ (\lambda + \mu) \frac{\partial v_j}{\partial x_j} - p \right\} + \mu \frac{\partial^2 v_i}{\partial x_j \, \partial x_j}. \tag{10.25}$$

This can also be written as

$$\nabla \cdot \underline{\underline{\sigma}} = \nabla \left\{ (\lambda + \mu) \nabla \cdot \underline{v} - p \right\} + \mu \Delta \underline{v}. \tag{10.26}$$

Using these expressions together with Eq. (10.3), one can obtain the following equivalent expressions of the Navier-Stokes equations:

$$\rho \left(\frac{\partial v_i}{\partial t} + v_j \frac{\partial v_i}{\partial x_j} \right) + \frac{\partial}{\partial x_i} \left\{ p - (\lambda + \mu) \nabla \cdot \underline{v} \right\} - \mu \Delta v_i = \rho b_i \tag{10.27}$$

and

$$\rho \left(\frac{\partial \underline{v}}{\partial t} + \underline{v} \cdot \nabla \underline{v} \right) + \nabla \left\{ p - (\lambda + \mu) \nabla \cdot \underline{v} \right\} - \mu \Delta \underline{v} = \rho \underline{b} \tag{10.28}$$

or, introducing the material derivative,

$$\rho \frac{D\underline{v}}{Dt} = \nabla \left\{ (\lambda + \mu) \nabla \cdot \underline{v} - p \right\} + \mu \Delta \underline{v} + \rho \underline{b}. \tag{10.29}$$

Another form of Eq. (10.29) is

$$\frac{D\rho \underline{v}}{Dt} + \rho \underline{v} \nabla \cdot \underline{v} = \nabla \left\{ (\lambda + \mu) \nabla \cdot \underline{v} - p \right\} + \mu \Delta \underline{v} + \rho \underline{b} \qquad (10.30)$$

which can be obtained by combining Eq. (10.29) with the equation of continuity, Eq. (10.2).

10.4 COMPLEMENTARY CONSTITUTIVE EQUATIONS

Pressure, density, and temperature are not independent; they are related by the equation of state of the fluid considered. There are a certain number of gases, whose behavior is closely mimicked by the *perfect gas equation*:

$$pV = nRT. \qquad (10.31)$$

Here v stands for the *specific volume*, which equals the reciprocal of the density. Thus, the equation

$$p = \rho nRT \qquad (10.32)$$

is frequently used instead of Eq. (10.31).

As for the heat flux, $\underline{q}$, occurring in the equation of energy balance, Eq. (10.7), the constitutive equation most frequently used in fluid mechanics is *Fourier's law for isotropic materials*:

$$\underline{q} = -k\nabla T. \qquad (10.33)$$

We recall that k is the *thermal conductivity*.

10.5 THE CONCEPTS OF INCOMPRESSIBLE AND INVISCID FLUIDS

Fluids occurring in physical reality are, to a greater or lesser extent, *viscid* and *compressible*. However, when studying them and modeling their behavior, the concepts of an *incompressible fluid* (but possibly *viscous*) and of an *inviscid fluid* (but possibly compressible) are very useful. The notion of an incompressible fluid allows for the development of models for a wide class of real fluids that are, in reality, slightly compressible, such as water, because it is found that the resulting equations satisfactorily mimic the behavior of slightly compressible fluids in many situations. In turn, the concept of an inviscid fluid yields models for a wide class of gases that are clearly compressible but only slightly *viscous*, such as air. Due to this fact, the theory of compressible but inviscid fluids is frequently called *gas dynamics*. In general, the theories presented here have a wide applicability when studying natural and industrial flows.

The relevance of compressibility depends not only on the nature of the fluid under consideration but also on the process that is being studied. In particular, the compressibility of a fluid rarely can be neglected when a fluid starts its motion, especially if such motion initiation is done in an abrupt manner. The compressibility can also not be neglected when the fluid moves at a velocity comparable to the speed of sound, measured with respect to objects immersed in the fluid.

A conspicuous feature of viscosity is that even when it is very small, it introduces a *singular perturbation* of the governing differential equations. When this happens, some of the conditions that should be fulfilled at the boundaries of solid objects in contact with the fluid cannot be satisfied when the viscosity is ignored. Thus, neglecting velocity produces boundary layers [7] in which the behavior predicted by the inviscid model is necessarily unsatisfactory.

10.6 INCOMPRESSIBLE FLUIDS

In this section we begin the study of *incompressible fluids* (generally with non-zero viscosity), and leave for later the study of compressible inviscid fluids. The assumption of incompressibility imposes a restriction on the motions that are admissible; this was discussed in Section 1.4.4 on page 14. Namely, the volume of the fluid must remain invariant during the motion of the fluid, and according to Eq. (1.40), this requires that particle velocities be divergence-free; that is, $\nabla \cdot \underline{v} = 0$. Motions that satisfy this equation are called *isochoric, solenoidal,* or *divergence-free*. In such a case, Eq. (10.22) reduces to

$$\underline{\underline{\sigma}} = -p\underline{\underline{I}} + 2\mu\underline{\underline{\varepsilon}} \tag{10.34}$$

and one has

$$\nabla \cdot \underline{\underline{\sigma}} = -\nabla p + \mu\Delta\underline{v}. \tag{10.35}$$

Then the *Navier-Stokes equation*, Eq. (10.28), takes the form

$$\left(\frac{\partial \underline{v}}{\partial t} + \underline{v} \cdot \nabla \underline{v} \right) = -\frac{1}{\rho}\nabla p + \upsilon\Delta\underline{v} + \underline{b}. \tag{10.36}$$

Here $\nu \equiv \rho^{-1}\mu$ is the *kinematic viscosity*. Sometimes, Eq. (10.36) is referred to as the *incompressible Navier-Stokes equation*.

Due to its importance, it is worth listing some other forms of the Navier-Stokes equation for incompressible fluids. They are:

$$\frac{\partial \underline{v}}{\partial t} + \nabla \frac{1}{2} \left(|\underline{v}|^2 \right) + \bar{\underline{\omega}} \times \underline{v} = -\frac{1}{\rho}\nabla p + \upsilon\Delta\underline{v} + \underline{b} \tag{10.37}$$

$$\frac{\partial \underline{v}}{\partial t} + \nabla \frac{1}{2} \left(|\underline{v}|^2 \right) + \bar{\underline{\omega}} \times \underline{v} = -\frac{1}{\rho}\nabla p - \upsilon\nabla\times\bar{\underline{\omega}} + \underline{b}. \tag{10.38}$$

Here $\bar{\underline{\omega}}$ is the *vorticity,* defined as

$$\bar{\underline{\omega}} \equiv \nabla \times \underline{v}. \tag{10.39}$$

Furthermore, we have used

$$\Delta\underline{v} = -\nabla \times \bar{\underline{\omega}}. \tag{10.40}$$

The concept of vorticity is very useful in many applications, and it is not difficult to derive some of the most basic equations that govern it from the above. Indeed, when the body forces are conservative, that is, when there is a function $\Phi\left(\underline{x}, t\right)$ such that

$$\underline{b} = -\nabla\Phi \tag{10.41}$$

then

$$\frac{\partial\bar{\omega}}{\partial t} + \nabla\times\left(\bar{\omega}\times\underline{v}\right) = v\Delta\bar{\omega} = -v\nabla\times\nabla\times\underline{\bar{\omega}}. \tag{10.42}$$

This is equivalent to

$$\frac{\partial\bar{\omega}}{\partial t} + \underline{v}\cdot\nabla\bar{\omega} = \bar{\omega}\cdot\nabla\underline{v} + v\Delta\bar{\omega}. \tag{10.43}$$

We observe that for compressible fluids, one has

$$\underline{\underline{\sigma}}:\nabla\underline{v} = p\nabla\cdot\underline{v} + 2\mu\underline{\underline{\varepsilon}}:\nabla\underline{v} = 2\mu\underline{\underline{\varepsilon}}:\nabla\underline{v}. \tag{10.44}$$

In addition,

$$\underline{\underline{\varepsilon}}:\nabla\underline{v} = \underline{\underline{\varepsilon}}:\left(\underline{\underline{\varepsilon}} + \underline{\underline{\omega}}\right) = \underline{\underline{\varepsilon}}:\underline{\underline{\varepsilon}}.$$

Therefore, because the product of a symmetric and a skew-symmetric matrix vanishes, the energy balance equation, Eq. (10.7), can be written as

$$\rho\frac{DE}{Dt} = k\Delta T + \rho h + 2\mu\underline{\underline{\varepsilon}}:\underline{\underline{\varepsilon}}. \tag{10.45}$$

10.7   INITIAL AND BOUNDARY CONDITIONS

Solutions to the system of equations discussed above constitute a complete model when certain initial and boundary conditions are prescribed. The initial conditions comprise the specification of the dependent variables at a certain time, usually referred to as the initial time, $t = 0$. As for the boundary conditions, generally when a viscous fluid is in contact with a solid, the fluid particles adhere to the surface of the body. This molecular phenomenon is a fact that has been verified experimentally, at least for a wide range of velocities of the fluid [10]. Because of this adherence, the velocity of the fluid relative to the surface of the solid is zero. In particular, if the body is at rest, then the velocity of the fluid is zero; since the velocity is a vector, this fact implies that both its normal and tangential components vanish.

On the other hand , if the fluid is inviscid, then the normal component has to be zero anyway, because of the impenetrability of the body, but the tangential component generally will be non-zero. When the tangential component is non-zero, one usually says that there is *slip*, while the condition of *adherence* is referred to as the *non-slip condition*. Some problems that have received special attention are flows past objects, such as the wing of an airplane, and fluid flow in a pipe. In the first, the region where the problem is formulated extends to infinity, and to get a well-posed problem, the asymptotic behavior of the solution there has to be prescribed. In both of them the non-slip condition has to be fulfilled in the solid surfaces, limiting the flow region.

When the changes of temperature are small, the fluid motion can be approximated as isothermal, in which case the energy balance equation is dropped out. Then the system of equations defined by the continuity and the Navier-Stokes equations constitute a complete model that when complemented by suitable initial and boundary conditions gives rise to well-posed problems. Putting together such equations, the following system, which governs the dynamics of incompressible fluids, is obtained:

$$\nabla \cdot \underline{v} = 0$$
$$\frac{\partial \underline{v}}{\partial t} + \underline{v} \cdot \nabla \underline{v} - \nu \Delta \underline{v} = -\rho^{-1} \nabla p + \underline{b}. \tag{10.46}$$

We observe that in this case, Eq. (10.2), the continuity equation, reduces to

$$\frac{D\rho}{Dt} = 0. \tag{10.47}$$

This equation is fulfilled if and only if each fluid particle conserves its density. In particular, if the density is uniform in space in the initial state, the density is constant throughout time and space.

10.8 VISCOUS INCOMPRESSIBLE FLUIDS: STEADY STATES

The study of steady-states for free-fluids is similar to the study of statics for solids. From Eq. (10.46) it is seen that for incompressible fluids, steady-states are governed by

$$\nabla \cdot \underline{v} = 0$$
$$\underline{v} \cdot \nabla \underline{v} - \nu \Delta \underline{v} = -\rho^{-1} \nabla p + \underline{b} \tag{10.48}$$

and well-posed problems are boundary-value problems; namely, the boundary conditions are

$$\underline{v}(\underline{x}) = 0 \text{ on } \partial \Omega. \tag{10.49}$$

This problem is very important, due to its practical applications. A conspicuous feature of it is that although the existence of a solution is granted under general conditions, its uniqueness can be shown only under very restrictive conditions [9]. This feature can be explained as follows: Generally, when a dissipative system starts its motion from a given initial state, subjected to certain boundary conditions, as time goes by it approaches a steady state. This explains why, under general conditions, the boundary-value problem possesses at least one solution. However, for a non-linear problem, such as the one considered here , nothing guarantees that the steady state that it approaches as time goes by is independent of the initial conditions.

10.9 LINEARIZED THEORY OF INCOMPRESSIBLE FLUIDS

Stokes equations are obtained as the linearized version of the incompressible Navier-Stokes equations for steady states. So when $\underline{v} = O(\varepsilon)$ and $\nabla \underline{v} = O(\varepsilon)$, $\underline{v} \cdot \nabla \underline{v} =$

$O\left(\varepsilon^2\right)$ and can be neglected in Eq. (10.48); one then gets

$$\nabla \cdot \underline{v} = 0$$
$$-\nu \Delta \underline{v} + \rho^{-1} \nabla p = \underline{b}. \tag{10.50}$$

In the literature describing numerical fluid dynamics, these equations are frequently written in a normalized form (that is, taking $\rho = 1$). Then one obtains

$$\nabla \cdot \underline{v} = 0$$
$$-\nu \Delta \underline{v} + \nabla p = \underline{b}. \tag{10.51}$$

10.10 IDEAL FLUIDS

The mathematical model of *ideal fluids* assumes that the fluid is incompressible and inviscid. In such a case, when the fluid is also homogeneous, the density of the fluid is a constant and the equation of continuity reduces to

$$\nabla \cdot \underline{v} = 0 \tag{10.52}$$

while Eqs. (10.29), with $\lambda = \mu = 0$, reduces to *Euler's equation*:

$$\frac{D\underline{v}}{Dt} = -\rho^{-1} \nabla p + \underline{b}. \tag{10.53}$$

The latter equation can also be written as

$$\frac{\partial \underline{v}}{\partial t} + \underline{v} \cdot \nabla \underline{v} = -\rho^{-1} \nabla p + \underline{b}. \tag{10.54}$$

Now

$$v_j \frac{\partial v_i}{\partial x_j} = v_j \frac{\partial v_j}{\partial x_i} + v_j \left(\frac{\partial v_i}{\partial x_j} - \frac{\partial v_j}{\partial x_i} \right) = \frac{1}{2} \frac{\partial v_j v_j}{\partial x_i} - v_j \left(\frac{\partial v_j}{\partial x_i} - \frac{\partial v_i}{\partial x_j} \right). \tag{10.55}$$

Hence, it can be verified that

$$\underline{v} \cdot \nabla \underline{v} = \frac{1}{2} \nabla |\underline{v}|^2 - \underline{v} \times \nabla \times \underline{v} \tag{10.56}$$

from which it follows that

$$\frac{D\underline{v}}{Dt} = \frac{\partial \underline{v}}{\partial t} + \frac{1}{2} \nabla |\underline{v}|^2 - \underline{v} \times \nabla \times \underline{v}. \tag{10.57}$$

Hence, Eq. (10.54) is

$$\frac{\partial \underline{v}}{\partial t} + \frac{1}{2} \nabla |\underline{v}|^2 - \underline{v} \times \nabla \times \underline{v} + \rho^{-1} \nabla p = \underline{b}. \tag{10.58}$$

In the absence of body forces, the scalar product of Eq. (10.58) and $\underline{v}$ yields *Bernoulli's equation*,

$$\underline{v} \cdot \frac{\partial \underline{v}}{\partial t} + \underline{v} \cdot \nabla \left(\frac{1}{2} |\underline{v}|^2 + \rho^{-1} p \right) = 0. \tag{10.59}$$

Therefore, an equation for an *ideal fluid* in the absence of body forces is

$$\frac{\partial |\underline{v}|^2}{\partial t} + \underline{v} \cdot \nabla \left(\frac{|\underline{v}|^2}{2} + \frac{p}{\rho} \right) = 0. \tag{10.60}$$

The envelopes of the particle-velocities field are called *streamlines*. The term $p_T \equiv p + \rho |\underline{v}|^2/2$ is called the *total pressure*. In particular, and in view of Eq. (10.59), in steady-state motions when $\partial \underline{v}/\partial t$ vanishes, the total pressure is constant along streamlines. This is a form of *Bernoulli's theorem*.

10.11 IRROTATIONAL FLOWS

The field, $\underline{v}(\underline{x}, t)$, of particle velocities is said to be *irrotational* when

$$\nabla \times \underline{v} = 0. \tag{10.61}$$

When the velocity field is irrotational, there exists a scalar function $\Phi(\underline{x}, t)$ such that

$$\underline{v} = \nabla \Phi. \tag{10.62}$$

Equations (10.52) and (10.62), together, imply that

$$\Delta \Phi = 0. \tag{10.63}$$

Hence,

$$\Delta \underline{v} = \Delta \nabla \Phi = \nabla \Delta \Phi = 0. \tag{10.64}$$

When this relation is substituted in the Navier-Stokes equation, Eq. (10.29), the latter equation reduces to Euler's equation, Eq. (10.53). However, this equation is equivalent to Eq. (10.58), which for irrotational flows and in the absence of body forces reduces to

$$\nabla \left(\frac{\partial \Phi}{\partial t} + \frac{1}{2} |\underline{v}|^2 + \rho^{-1} \nabla p \right) = 0. \tag{10.65}$$

This is tantamount to the condition that the expression occurring on the left hand of Eq. (10.65) be a function of time, exclusively; that is,

$$\frac{\partial \Phi}{\partial t} + \frac{1}{2} |\underline{v}|^2 + \rho^{-1} p = F(t). \tag{10.66}$$

Thus, we have shown that irrotational flows of incompressible fluids satisfy the *Navier-Stokes equation* if, and only if, Eq. (10.66) is satisfied. In particular, for steady states, Eq. (10.66) is

$$p_T \equiv p + \rho |\underline{v}|^2/2 = const. \tag{10.67}$$

This implies that

$$p = p_T - \rho |\underline{v}|^2 / 2. \tag{10.68}$$

It is not difficult to extend Eq. (10.66) to include non-vanishing body forces, provided that they can be derived from a potential. Indeed, if such a body force is given by

$$\underline{b} = \nabla \psi \tag{10.69}$$

then a slight modification of the arguments that led to Eq. (10.66) yields

$$\frac{\partial \Phi}{\partial t} + \frac{1}{2} |\underline{v}|^2 + \rho^{-1} p + \psi = F(t). \tag{10.70}$$

10.12   EXTENSION OF BERNOULLI'S RELATIONS TO COMPRESSIBLE FLUIDS

We would like to obtain relations similar to Eq. (10.66) that hold for a compressible fluid. To this end, we assume that the pressure is given by a thermodynamic relation of the form

$$p = p(\rho, s). \tag{10.71}$$

Here s is the *entropy*. Furthermore, it will be assumed that the flow is *homentropic*, in which case the pressure is a function of the density, exclusively:

$$p = p(\rho). \tag{10.72}$$

In such a case, the thermodynamic relation for the enthalpy

$$dH = T \, ds + \frac{dp}{\rho} \tag{10.73}$$

reduces to

$$dH = \frac{dp}{\rho}. \tag{10.74}$$

Therefore,

$$\nabla H = \frac{\nabla p}{\rho} \tag{10.75}$$

and the new version of Eq. (10.65) is

$$\nabla \left(\frac{\partial \Phi}{\partial t} + \frac{1}{2} |\underline{v}|^2 + \psi + H \right) = 0. \tag{10.76}$$

Hence

$$\frac{\partial \Phi}{\partial t} + \frac{1}{2} |\underline{v}|^2 + \psi + H = F(t). \tag{10.77}$$

10.13 SHALLOW-WATER THEORY

Shallow-water theory is, in its lowest approximation, the basic theory used in hydraulics by engineers when dealing with flows in open channels; it is sometimes referred to as the theory of long waves. This theory applies when the domain occupied by the fluid has horizontal dimensions which are much larger than its vertical thickness. To be specific, taking x_3 in the vertical direction and the other two coordinates in the horizontal plane, it is assumed that such a domain is limited at the top and bottom by

$$\eta\left(x_1, x_2, t\right) - x_3 = 0$$
$$h\left(x_1, x_2\right) + x_3 = 0 \tag{10.78}$$

respectively. It is understood that the top is a *free surface*; by this we mean that the fluid particles there are subjected to atmospheric pressure (which will be taken to be zero in what follows). In particular, $h\left(x_1, x_2\right)$ is a given function while $\eta\left(x_1, x_2, t\right)$ is not and has to be obtained as part of the solution of the problem.

If we define $F\left(\underline{x}, t\right) \equiv \eta\left(x_1, x_2, t\right) - x_3$ and $G\left(\underline{x}, t\right) \equiv h\left(x_1, x_2\right) + x_3$, then Eq. (10.78) is

$$F\left(\underline{x}, t\right) = 0 \text{ and } G\left(\underline{x}, t\right) = 0. \tag{10.79}$$

A basic assumption in fluid mechanics and in many branches of continuous mechanics is that any particle of a fluid body that is once at the boundary of such a body remains on the boundary. This assumption, together with Eq. (10.79), implies that

$$\frac{DF}{Dt}\left(\underline{x}, t\right) = 0 \text{ whenever } x_3 = \eta\left(x_1, x_2, t\right)$$
$$\frac{DG}{Dt}\left(\underline{x}, t\right) = 0 \text{ whenever } x_3 = -h\left(x_1, x_2\right). \tag{10.80}$$

On the other hand, the fluid will be taken to be an ideal fluid; that is, it is incompressible and inviscid. Hence, the motion of the fluid is governed by Eqs. (10.52) and (10.54).

Due to the different roles played by the vertical and horizontal dimensions, we write

$$\underline{v} \equiv \underline{u} + \underline{w} \tag{10.81}$$

where $\underline{u}$ and $\underline{w}$ are the horizontal and vertical components of the velocity, respectively. As for the vertical coordinate, we write z for it and assume that it grows upward.

Using this notation, it is seen that Eqs. (10.80) are

$$\eta_t + \underline{u} \cdot \nabla\eta - w = 0 \text{ and } \underline{u} \cdot \nabla h + w = 0 \tag{10.82}$$

at the top and bottom of the fluid body, respectively. As for Eqs. (10.52) and (10.54), they are

$$\nabla \cdot \underline{u} + \frac{\partial w}{\partial z} = 0 \tag{10.83}$$

and

$$\frac{\partial \underline{u}}{\partial t} + \underline{u} \cdot \nabla \underline{u} + w \frac{\partial \underline{u}}{\partial z} = -\rho^{-1} \nabla p$$

$$\frac{\partial w}{\partial t} + \underline{u} \cdot \nabla w + w \frac{\partial w}{\partial z} = -\rho^{-1} \frac{\partial p}{\partial z} - g \tag{10.84}$$

respectively. Here, g is the magnitude of the acceleration due to gravity.

The shallow-water approximation is obtained when it is assumed that the left-hand member of the second of Eqs. (10.84), as well as the term $w \, \partial \underline{u}/\partial z$, can be neglected, and we seek a horizontal velocity field $\underline{u}(x_1, x_2, t)$, which is independent of the vertical coordinate z. Then, we have

$$\frac{\partial \underline{u}}{\partial t} + \underline{u} \cdot \nabla \underline{u} + \rho^{-1} \nabla p = 0$$

$$\frac{\partial p}{\partial z} + \rho g = 0. \tag{10.85}$$

Integration of the second of these equations yields

$$p = \rho g \left(\eta - z \right). \tag{10.86}$$

When this result is used in Eq. (10.85), one gets

$$\frac{\partial \underline{u}}{\partial t} + \underline{u} \cdot \nabla \underline{u} + g \nabla \eta = 0. \tag{10.87}$$

Integrating Eq. (10.83) in the vertical direction and using the fact that $\underline{u}(x_1, x_2, t)$ is independent of z, one obtains

$$(\eta + h) \nabla \cdot \underline{u} + w \left(\underline{x}, \eta, t \right) - w \left(\underline{x}, -h, t \right) = 0. \tag{10.88}$$

Using this result together with Eq. (10.82), it can be verified that

$$\nabla \cdot \{(\eta + h) \underline{u}\} + \eta_t \left(\underline{x}, t \right) = 0. \tag{10.89}$$

Equations (10.87) and (10.89), together, are the basic differential equations of shallow water theory.

In order to obtain some insight into the behavior of the shallow-water model just derived, consider a linearized version of it. Assume that both $\underline{u}$ and η, together with their derivatives, are small and neglect second-order terms in Eqs. (10.87) and (10.89). Then we get

$$\frac{\partial \underline{u}}{\partial t} + g \nabla \eta = 0 \tag{10.90}$$

and

$$\nabla \cdot (h \underline{u}) + \eta_t \left(\underline{x}, t \right) = 0. \tag{10.91}$$

Multiplying Eqs. (10.90) by h and taking the divergence of the resulting equation, one obtains

$$\frac{\partial}{\partial t} \nabla \cdot (h \underline{u}) + g \nabla \cdot (h \nabla \eta) = 0. \tag{10.92}$$

On the other hand, taking the derivative with respect to time of Eq. (10.91), we get

$$\frac{\partial}{\partial t} \nabla \cdot (h\underline{u}) + \eta_{tt}(\underline{x}, t) = 0. \tag{10.93}$$

Subtracting Eq. (10.92) from Eq. (10.93), a hyperbolic equation for η is derived:

$$\eta_{tt}(\underline{x}, t) - g\nabla \cdot (h\nabla\eta) = 0. \tag{10.94}$$

In particular, when h is a constant, this equation can be written as

$$\frac{1}{gh}\eta_{tt}(\underline{x}, t) - \Delta\eta = 0. \tag{10.95}$$

This is the *wave equation with wave speed equal to* $\sqrt{gh}$. As a matter of fact, it can be shown that $\sqrt{gh}$ is indeed the speed of propagation of waves in models governed by the shallow-water theory.

10.14   INVISCID COMPRESSIBLE FLUIDS

The motion of compressible fluids, which are inviscid, is governed by the equation of continuity:

$$\frac{\partial \rho}{\partial t} + \nabla \cdot (\rho\underline{v}) = 0 \tag{10.96}$$

together with the balance of linear momentum. This for inviscid fluids reduces to

$$\rho\frac{D\underline{v}}{Dt} + \nabla p = \rho\underline{b} \tag{10.97}$$

or, more explicitly, we can write

$$\rho\left(\frac{\partial\underline{v}}{\partial t} + \underline{v} \cdot \nabla\underline{v}\right) + \nabla p = \rho\underline{b}. \tag{10.98}$$

We present in what follows some salient features of compressible fluids, neglecting the effect of viscosity. Some of the most conspicuous features that distinguish the behavior of compressible fluids from that of incompressible fluids are:

1. The speed of transmission of signals by an incompressible fluid is infinite, while it is finite for a compressible fluid;

2. The flow initiation for a compressible fluid is drastically different from that for an incompressible fluid; and

3. The governing equations of inviscid compressible fluids admit discontinuous solutions, and shock formation is possible, whereas those of incompressible fluids do not admit discontinuous solutions.

We illustrate these differences in what follows.

10.14.1 Small perturbations in a compressible fluid: the theory of sound

The continuity equation, Eq. (10.96), can be written as

$$\frac{\partial \rho}{\partial t} + \underline{v} \cdot \nabla \rho + \rho \nabla \cdot \underline{v} = 0. \tag{10.99}$$

A linearized version of it, when $\underline{v}$, $\nabla \rho$, $\nabla \cdot \underline{v}$, and $\rho - \rho_0$ are $O(\varepsilon)$ and terms $O(\varepsilon^2)$ are neglected, is

$$\frac{\partial \rho}{\partial t} + \rho_0 \nabla \cdot \underline{v} = 0. \tag{10.100}$$

Similarly, a linearized version of Navier-Stokes equation, Eq. (10.98), is

$$\rho_0 \frac{\partial \underline{v}}{\partial t} + \nabla p = 0. \tag{10.101}$$

Under the assumption that

$$p = p(\rho) \tag{10.102}$$

Eq. (10.100) can be written as

$$\frac{1}{p'(\rho_0)} \frac{\partial p}{\partial t} + \rho_0 \nabla \cdot \underline{v} = 0. \tag{10.103}$$

When writing this equation, we have neglected terms $O(\varepsilon^2)$. Taking the derivative with respect to time, Eq. (10.103) yields

$$\frac{1}{p'(\rho_0)} \frac{\partial^2 p}{\partial t^2} + \rho_0 \nabla \cdot \frac{\partial \underline{v}}{\partial t} = 0. \tag{10.104}$$

On the other hand, applying the divergence operator to Eq. (10.101), one obtains

$$\rho_0 \nabla \cdot \frac{\partial \underline{v}}{\partial t} + \Delta p = 0. \tag{10.105}$$

Equations (10.104) and (10.105), together, imply that

$$\frac{1}{a^2} \frac{\partial^2 p}{\partial t^2} - \Delta p = 0. \tag{10.106}$$

Equation (10.106) is the wave equation with *sound speed*, a, equal to

$$a(\rho_0) = \sqrt{p'(\rho_0)}. \tag{10.107}$$

Here the equations of motion were linearized; so Eq. (10.106) is a linear equation and in this case the *speed of sound* is independent of the pressure, p. However, in what follows we deal with Eqs. (10.96) to (10.98) before linearization, when the sound speed is, in general, a function of the pressure or the density.

10.14.2 Initiation of motion

Consider a straight pipe filled with a fluid, closed at one end by a piston. Next, let the piston be pushed into the pipe or withdrawn from it. We now study the motion of the fluid produced in this manner, under the assumption that the velocity is uniform in each cross section of the pipe. Thus, the physical space of our model involves only one dimension.

The one-dimensional versions of Eqs. (10.96) and (10.98) without body forces are

$$
\begin{aligned}
\frac{\partial \rho}{\partial t} + v\frac{\partial \rho}{\partial x} + \rho\frac{\partial v}{\partial x} &= 0 \\
\frac{\partial v}{\partial t} + v\frac{\partial v}{\partial x} + \frac{1}{\rho}\frac{\partial p}{\partial x} &= 0
\end{aligned}
\tag{10.108}
$$

Let us now define the function w by

$$
\frac{dw}{d\rho} \equiv \frac{a}{\rho}.
\tag{10.109}
$$

Then expressing ρ and p in terms of w, Eqs. (10.108) become

$$
\begin{aligned}
\frac{\partial w}{\partial t} + v\frac{\partial w}{\partial x} + a\frac{\partial v}{\partial x} &= 0 \\
\frac{\partial v}{\partial t} + v\frac{\partial v}{\partial x} + a\frac{\partial w}{\partial x} &= 0.
\end{aligned}
\tag{10.110}
$$

Now define

$$
q \equiv \frac{1}{2}(w+v) \text{ and } r \equiv \frac{1}{2}(w-v).
\tag{10.111}
$$

Then, adding and subtracting the pair of Eqs. (10.110), it is seen that

$$
\begin{aligned}
\frac{\partial q}{\partial t} + (v+a)\frac{\partial q}{\partial x} &= 0 \\
\frac{\partial r}{\partial t} + (v-a)\frac{\partial r}{\partial x} &= 0.
\end{aligned}
\tag{10.112}
$$

These are Riemann's equations [6]. They state that

$$
\begin{aligned}
\frac{dq}{dt} &= 0 \text{ along } \frac{dx}{dt} = v+a \\
\frac{dr}{dt} &= 0 \text{ along } \frac{dx}{dt} = v-a.
\end{aligned}
\tag{10.113}
$$

Lines satisfying $dx/dt = v+a$ are said to be *advancing Mach lines,* while those that satisfy $dx/dt = v-a$ are *receding Mach lines.* Furthermore, a subdomain where q, or r, is a constant is called a *simple wave.*

In the piston problem, every point of the fluid domain (Fig. 10.1) is connected to the positive x-*axis* by a receding Mach line. The initial condition there is that the gas is at rest and in a uniform thermodynamic state; that is,

$$
\left.\begin{aligned}
v &= 0 \\
w &= \text{const} = w_0
\end{aligned}\right\} \Rightarrow \{q = r = w_0/2.
\tag{10.114}
$$

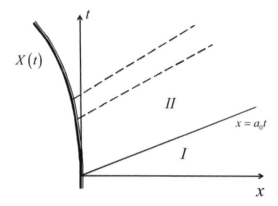

Figure 10.1   Fluid domain for the piston problem with fluid at rest.

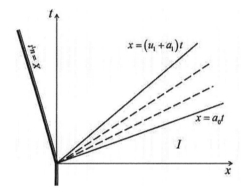

Figure 10.2   Fluid domain for the piston problem with piston receding.

Thus, the space-time domain of the problem must consist of *simple waves* and possibly, uniform subdomains. On the advancing Mach lines, both q and r have to be constants. Hence, so are v and a. This shows that the advancing Mach lines are straight lines.

In Fig. 10.2 the piston recedes at a certain velocity, possibly non-uniform. Then we can distinguish three regions of the x-t plane. Region I is a uniform one, where all variables are constant and the fluid particles are at rest. Region II is occupied by simple waves, made of straight Mach lines that converge at the origin of the x-t plane. Finally, region III is also made of simple waves, but they are not converging. If the piston recedes at a uniform velocity in Fig. 10.2, region III is also uniform, but the fluid particles are not at rest there; indeed, they move with the velocity of the piston.

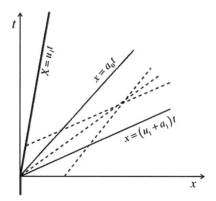

Figure 10.3 Fluid domain for the piston problem with piston advancingt.

In the cases illustrated in Figs. 10.1 and 10.2, the construction above supplies a solution to the boundary value problem associated with the piston example. The situation illustrated in Fig. 10.2 is quite different. There, the piston advances into the region initially occupied by the fluid at rest. Let the equation

$$x = X(t); \; X(t) \geq 0 \tag{10.115}$$

describe the motion of the piston. Then, if $X''(t) > 0$, the advancing Mach lines converge as time increases, creating an envelope after a certain time. Moreover, each of these Mach lines carries its own value of q, giving rise to an incompatible (multiple-valued) solution and unbounded gradients. This shows that the problem, as formulated above, does not possess a solution in this case, and the whole concept of what a solution is has to be revised.

10.14.3 Discontinuous models and shock conditions

As was discussed in Chapter 2, every model of a one-phase system consists of a family of intensive properties $\left\{ \psi^1, ..., \psi^N \right\}$ that, during some time interval, satisfy the balance conditions:

$$\left. \begin{array}{l} \dfrac{\partial \psi^\alpha}{\partial t} + \nabla \cdot (\underline{v}\psi^\alpha) = \nabla \cdot \underline{\tau}^\alpha + g^\alpha \text{ in } \Omega - \Sigma \\[2mm] [\![\psi^\alpha (\underline{v} - \underline{v}_\Sigma) - \underline{\tau}^\alpha]\!] \cdot \underline{n} = 0 \text{ on } \Sigma \end{array} \right\}, \; \alpha = 1, ..., N. \tag{10.116}$$

Thus, in general, a discontinuous model consists of a domain Ω, a surface Σ, and a family of functions $\left\{ \psi^1, ..., \psi^N \right\}$, which together with their derivatives may have jump discontinuities across Σ. Furthermore, the family of functions satisfy the balance conditions of Eq. (10.116), which consist of partial differential equations and jump conditions.

In the case of compressible inviscid fluids, the family of intensive properties are the *density*, ρ; the *density of linear momentum*, $\rho\underline{v}$; the *density of kinetic energy*, $\rho|\underline{v}|^2/2$; and the density of *internal energy*, ρE. The differential equations have been discussed in previous sections, and the jump conditions, which in this context are usually called *shock conditions*, were discussed in Sections 2.4 to 2.9. Here we recall the main results. The shock conditions for the density are

$$[\![\rho(\underline{v} - \underline{v}_\Sigma)]\!] \cdot \underline{n} = 0 \text{ on } \Sigma. \tag{10.117}$$

This equation states that $\rho(\underline{v} - \underline{v}_\Sigma) \cdot \underline{n}$ is continuous across Σ. Taking this into account, the jump condition for the density of linear momentum reduces to

$$[\![\underline{v}]\!]\,\rho(\underline{v} - \underline{v}_\Sigma) \cdot \underline{n} - [\![\underline{\underline{\sigma}}]\!] \cdot \underline{n} = 0 \text{ on } \Sigma. \tag{10.118}$$

Recalling that for an inviscid fluid Eq. (10.13) reduces to

$$\underline{\underline{\sigma}} = -p\underline{\underline{I}} \tag{10.119}$$

we obtain

$$[\![\underline{v}]\!]\,\rho(\underline{v} - \underline{v}_\Sigma) \cdot \underline{n} + [\![p]\!]\,\underline{n} = 0 \text{ on } \Sigma. \tag{10.120}$$

Decomposing the fluid velocity into its normal and tangential components we obtain

$$\underline{v} = v_n\underline{n} + v_t\underline{t}. \tag{10.121}$$

Here $\underline{t}$ is a unit vector in the direction of the tangential (to Σ) component of $\underline{v}$. Then Eq. (10.120) is equivalent to the following two conditions:

$$
\begin{aligned}
[\![v_t]\!] &= 0 \\
[\![p]\!] &= -[\![v_n]\!]\,\rho(\underline{v} - \underline{v}_\Sigma) \cdot \underline{n}.
\end{aligned}
\tag{10.122}
$$

Thus, one can say that the velocity component tangential to the shock is not affected by the shock, since it is continuous. The other inequality in Eq. (10.122) relates to the jump of the pressure with the jump of the normal component of the velocity. The shock relations for the total energy are

$$\left[\!\left[\frac{1}{2}\rho v^2(\underline{v} - \underline{v}_\Sigma) + p\underline{v}\right]\!\right] \cdot \underline{n} + [\![E]\!]\,\rho(\underline{v} - \underline{v}_\Sigma) \cdot \underline{n} + [\![\underline{q}]\!] \cdot \underline{n} = 0 \tag{10.123}$$

but

$$
\begin{aligned}
\left[\!\left[\frac{1}{2}\rho v^2(\underline{v} - \underline{v}_\Sigma)\right]\!\right] \cdot \underline{n} \\
= \frac{1}{2}[\![v^2]\!]\,\rho(\underline{v} - \underline{v}_\Sigma) \cdot \underline{n} \\
= \frac{1}{2}[\![v_n^2]\!]\,\rho(\underline{v} - \underline{v}_\Sigma) \cdot \underline{n} \\
= \dot{v}_n[\![v_n]\!]\,\rho(\underline{v} - \underline{v}_\Sigma) \cdot \underline{n} \\
= -\dot{v}_n[\![p]\!]'.
\end{aligned}
\tag{10.124}
$$

On the other hand,

$$[[p\underline{v}]] \cdot \underline{n} = [[pv_n]] = \dot{v}_n [[p]] + \dot{p} [[v_n]]. \tag{10.125}$$

Therefore,

$$\left[\left[\frac{1}{2}\rho v^2 (\underline{v} - \underline{v}_\Sigma) + p\underline{v}\right]\right] \cdot \underline{n} = \dot{p} [[v_n]]. \tag{10.126}$$

Hence,

$$[[E]] \rho (\underline{v} - \underline{v}_\Sigma) \cdot \underline{n} + [[q]] \cdot \underline{n} + \dot{p} [[v_n]] = 0. \tag{10.127}$$

If there is no heat flux, $[[q]] = 0$, Eq. (10.127) reduces to

$$[[E]] \rho (\underline{v} - \underline{v}_\Sigma) \cdot \underline{n} + \dot{p} [[v_n]] = 0. \tag{10.128}$$

10.15 SUMMARY

In this chapter we consider the fundamental equations of fluid mechanics. We begin with a review of the balance equations of mass, energy, and momentum. Next the constitutive equations for Newtonian fluids are presented. In the following section we show how one uses the information above to formulate the Navier-Stokes equations. The complementary constitutive equations of state are discussed next. We then turn to the concepts of incompressible and inviscid fluids. Boundary and initial conditions relevant to the equations above are introduced next. The following sections address viscous incompressible fluids, beginning with steady-state conditions. There follows a brief consideration of ideal fluids and irrotational flows. An extension of Bernoulli's relations for a compressible fluid follows. A discussion of shallow-water theory is followed by a rather extensive consideration of inviscid compressible flows. The chapter closes with a simple illustration of shock formation and the application to shocks in compressible fluid dynamics of the general jump conditions.

EXERCISES

10.1 Show that the balance of linear momentum can also be expressed as

$$\frac{D\rho\underline{v}}{Dt} + \rho\underline{v}\nabla \cdot \underline{v} = \nabla \cdot \underline{\underline{\sigma}} + \rho\underline{b}. \tag{10.129}$$

10.2 State the first Stokes' axiom and show that it implies that when the fluid is at rest:

  a) At any surface whose normal vector is $\underline{n}$, the traction is given by

$$\underline{T} = -p\underline{n}. \tag{10.130}$$

b) The stress tensor is

$$\underline{\underline{\sigma}} = -p\underline{\underline{I}}. \tag{10.131}$$

10.3 Define the viscous stress tensor. Give the expression of the stress tensor for an inviscid fluid.

10.4 Show that the set of 3×3 symmetric matrices constitute a linear space of dimension 6. Furthermore, each of Eqs. (9.24) and (10.16) defines linear transformations of such a linear space into itself. These two linear transformations are similar, in many respects, especially when the elastic material considered is isotropic, since in such a case Eqs. (9.82) and (10.17) hold simultaneously. For example, in Section 9.7 it was shown that a complete set of proper states is constituted by the set of isotropic states of strain and the set of isochoric states of strain. The corresponding proper values are $3k$ and 2μ, with multiplicities 1 and 5, respectively. In this exercise the reader is required to find a complete set of proper states for the linear transformation of (10.16), their corresponding proper values, and the multiplicity of each of them.

10.5 In spite of the similarities between the elastic tensor and the matrix of viscous coefficients discussed in exercise 10.4, there are sharp differences in the physical interpretations of one and the other. Indeed, since the viscous coefficients are given by Eq. (10.17), the following symmetries are enjoyed by both the elastic tensor and the viscous coefficients:

$$C_{ijpq} = C_{jipq} \text{ and } c_{ijpq} = c_{jipq} \tag{10.132}$$

$$C_{jipq} = C_{ijqp} \text{ and } c_{ijpq} = c_{ijqp} \tag{10.133}$$

and

$$C_{ijpq} = C_{pqij} \text{ and } c_{ijpq} = c_{pqij}. \tag{10.134}$$

Show that the first two symmetries have similar physical meanings.

10.6 First, for the case of a linearly elastic material and using Eq. (9.29), show that the symmetry of Eq. (10.134) is tantamount to the condition that the total work done when deforming such a material vanishes whenever it is returned to its initial state– see, Eq. (9.29)-. Is it possible to establish a similar result for the case of a viscous fluid? If your answer is negative, does this mean that in the motion of a viscous fluid there is dissipation of energy while the motion of an elastic solid is conservative (i.e., there is no dissipation of energy)? The use of the balance of energy, in the form of Eq.(2.66), will help to enlighten this point.

10.7 Show that Eqs. (10.29) and (10.30) are equivalent. Then derive the Navier-Stokes equation in the form of Eq. (10.29).

10.8 Show that an for an ideal gas, when isothermal flow is assumed, Eq. (10.106) can also be written in terms of the fluid density as

$$\frac{1}{a^2}\frac{\partial^2 \rho}{\partial t^2} - \Delta\rho = 0. \tag{10.135}$$

10.9   Write Eq. (10.135) for the case when ρ is a function of one of the spatial coordinates and time, exclusively. Show that in that case, any solution of Eq. (10.135) can be written as

$$\rho(x,t) = f(x+at) + g(x-at) \tag{10.136}$$

where f and g are functions of only one variable. Physically, this manner of expressing $\rho(x,t)$ can be interpreted as the superposition of an advancing and a receding wave. Which of them is the advancing wave and which is the receding wave?

10.10   In this exercise we consider an incompressible fluid in two dimensions. Show that in such a case:

a) There is function ψ, called the *stream function*, such that

$$\nabla\psi = \begin{pmatrix} -v_y \\ v_x \end{pmatrix}. \tag{10.137}$$

Here we have written v_x and v_y for the velocity components.

b) The stream function is uniquely defined by Eq. (10.137), except for a constant function.

c) When the flow is irrotational, the stream function is a harmonic function; i.e.,

$$\Delta\psi = 0. \tag{10.138}$$

REFERENCES

1. Donaldson, E.C., G.V. Chiligarian, and T.F. Yen, *Enhanced Oil Recovery, Vol.I: Fundamentals and Analysis*, Developmments in Petroleum Science, Vol.17A, Elsevier, Amsterdam, 1985.

2. Emanuel, G., *Analytical Fluid Dynamics,* CRC Press, Boca Raton, FL, 2001.

3. Garabedian, P.R., *Partial Differential Equations*, Wiley, New York, 1964.

4. Glowinski, R., Numerical Methods for Fluids*, Handbook of Numerical Analysis, Part 3,* Eds. P.G. Ciarlet and J.L. Lions, North-Holland, Amsterdam, 2003.

5. Granger, R.A., *Fluid Mechanics,* Dover, New York, 1995.

6. Meyer, R.E., *Introduction to Mathematical Fluid Dynamics*, Dover, New York, 1982 .

7. Schlichting, H., *Boundary Layer Theory,* McGraw-Hill, New York, 1960.

8. Stoker, J.J., *Water Waves*, Interscience, New York, 1957.

9. Temam, R., *Navier-Stokes Equations: Theory and Numerical Analysis*, American Mathematical Society, Providence, Rhode Island, 1984.

10. Warsi, Z.U.A., *Fluid Dynamics: Theoretical and Computational Approaches*, 2nd Ed., CRC Press, Boca Raton, FL,1998.

11. Yeh, H. and J.L. Abrams, , *Principles of Mechanics of Solids and Fluids,* Vol. 1: Particle and Rigid-Body Mechanics, McGraw-Hill, New York, 1960.

APPENDIX A

PARTIAL DIFFERENTIAL EQUATIONS

As noted in the text, continuous models are described by partial differential equations. Thus, the subject matter of this appendix is fundamental to the mathematical and computational modeling of continuous systems, and readers are encouraged to study additional references on the topic and associated numerical methods. To assist in this activity, a good number of books are listed in the bibliography of Chapter 2 this appendix . The main thrust of this appendix is the study of well-posed problems for the various types of partial differential equations.

A.1 CLASSIFICATION

In this appendix we study three types of partial differential equations: *elliptic*, *parabolic,* and *hyperbolic* equations. These three classes of equations include the partial differential equations most frequently encountered in engineering and science. In particular, practically all the continuous models studied in this book are governed by equations and systems of equations that belong to one of the three types mentioned above. It should be mentioned that non-diffusive transport of solutes by a fluid gives rise to first-order equations, which are usually considered to be hyperbolic [1],[2].

Mathematical Modeling in Science and Engineering: An Axiomatic Approach.
By Ismael Herrera and George F. Pinder Copyright © 2012 John Wiley & Sons, Inc.

An advantage of proceeding in this manner is that many properties are common to each member of the same class. We will place special emphasis on *linear equations* because they play a central role in mathematical and computational modeling of continuous systems. They possess very convenient algebraic and approximating properties, which permit their widespread use for solving very complicated non-linear problems.

A domain Ω of R^n will be considered; here n is the number of independent variables of the function u. Every second-order partial differential equation can be written as

$$Lu \equiv -\nabla \cdot (\underline{a} \cdot \nabla u) + \nabla \cdot (\underline{b}u) + cu = f_\Omega. \tag{A.1}$$

Here for each $\underline{x} \in \Omega$, $\underline{a}(\underline{x})$ is a symmetric matrix, $\underline{b}(\underline{x})$ is a vector, and $c(\underline{x})$ is a real number (a scalar).

The *linear differential operator* L so defined is the most general second-order linear differential operator. If the elements of matrix $\underline{a}$ and vector $\underline{b}$ are denoted by a_{ij} and b_i, respectively, then Eq. (A.1) can be written as

$$\sum_{i=1}^{n}\sum_{j=1}^{n} \frac{\partial}{\partial x_j}\left(a_{ij}\frac{\partial u}{\partial x_j}\right) + \sum_{j=1}^{n}\frac{\partial}{\partial x_j}(b_j u) + cu = f_\Omega. \tag{A.2}$$

The *eigenvectors* of the matrix $\underline{a}(\underline{x})$ constitute a *basis* of R^n, since it is symmetric, and the differential operator L is said to be:

1. *Elliptic* when all the *proper values (eigenvalues)* are non-zero and of the same sign, in which case a is *definite* (either positive or negative) ;

2. *Hyperbolic* when all the proper-values are non-zero and one of them is opposite in sign to all the others; and

3. *Parabolic* when all the proper-values except one are non-zero and all those that are non-zero have the same sign.

When the number of independent variables is two, these conditions can be expressed as

$$\text{I. } Elliptic \Leftrightarrow a_{12}^2 - a_{11}a_{22} < 0$$
$$\text{II. } Hyperbolic \Leftrightarrow a_{12}^2 - a_{11}a_{22} > 0$$
$$\text{III. } Parabolic \Leftrightarrow a_{12}^2 - a_{11}a_{22} = 0.$$

A notation that is commonly used for this case is $a \equiv a_{11}$, $b \equiv a_{12} = a_{21}$, and $c \equiv a_{22}$, in which case

$$\text{I. } Elliptic \Leftrightarrow b^2 - ac < 0$$
$$\text{II. } Hyperbolic \Leftrightarrow b^2 - ac > 0$$
$$\text{III. } Parabolic \Leftrightarrow b^2 - ac = 0.$$

A.2   CANONICAL FORMS

Next we present three equations with constant coefficients, which constitute proto-types of each of the different classes of partial differential equations. These equations illustrate effectively their properties and their behaviors. For elliptic equations, such a prototype is the Laplace equation, defined to be

$$\sum_{i=1}^{n} \frac{\partial^2 u}{\partial x_i^2} = 0. \tag{A.3}$$

In the case of hyperbolic and parabolic equations it is customary to single out one of the independent variables and associate it with time. Such a notation is motivated by the fact that in the physical system that the equation mimics, that independent variable corresponds to time. For example, the *wave equation*, which is the prototype of hyperbolic equations, is written as

$$\frac{\partial^2 u}{\partial t^2} - \sum_{i=1}^{n-1} \frac{\partial^2 u}{\partial x_i^2} = 0. \tag{A.4}$$

or, introducing the *Laplacian* operator

$$\frac{\partial^2 u}{\partial t^2} - \Delta u = 0. \tag{A.5}$$

Similarly, the *heat equation*, which is the prototype of parabolic equations, is written as

$$\frac{\partial u}{\partial t} - \sum_{i=1}^{n-1} \frac{\partial^2 u}{\partial x_i^2} \tag{A.6}$$

or

$$\frac{\partial u}{\partial t} - \Delta u = 0. \tag{A.7}$$

A.3   WELL-POSED PROBLEMS

Generally, each partial differential equation (or system of such equations) has many solutions and in order to formulate problems that have only one solution, it is nec-essary to supplement the equations with suitable boundary and possibly initial con-ditions. Such a problem is said to be \well-posed" when it possesses one and only one solution, which depends continuously on the boundary and initial conditions. A problem that involves a partial differential equation and boundary conditions is said to be a boundary-value problem (or an *initial-boundary-value problem* when it also involves initial conditions). The study and analysis of well-posed problems for partial differential equations have been the subject of extensive classical research. For linear equations very comprehensive results are available that have been collected and summarized in various books.

A.3.1   Boundary-value problems: the elliptic case

For the Laplace differential operator we consider a domain Ω, contained in R^n and with boundary $\partial\Omega$. Then the most general boundary-value problem is: Given the functions f_Ω and g_∂ (f_Ω defined in Ω and g_∂ defined in $\partial\Omega$), find a function u such that

$$
\Delta u = f_\Omega \text{ in } \Omega
$$
$$
\alpha\frac{\partial u}{\partial n} + \beta u = 0 \text{ on } \partial\Omega.
$$
(A.8)

Here it is assumed that α and β are real numbers such that $\alpha^2 + \beta^2 = 1$. Such boundary conditions are said to be Robin boundary conditions. There are particular cases of this general problem that occur in many instances and have received much attention. When $\alpha = 0$, one obtains the Dirichlet problem:

$$
\Delta u = f_\Omega \text{ in } \Omega
$$
$$
u = 0 \text{ on } \partial\Omega
$$
(A.9)

and when $\beta = 0$, one obtains the Neumann problem:

$$
\Delta u = f_\Omega \text{ in } \Omega
$$
$$
\frac{\partial u}{\partial n} = 0 \text{ on } \partial\Omega.
$$
(A.10)

A.3.2   Initial-boundary-value problems

To formulate this kind of problem, generally one considers the domain $\Omega \times [0, T]$, with $T > 0$, where $[0, T]$ is an interval of the real line. In some cases, usually of theoretical interest, one may treat the whole infinite interval $[0, \infty)$.

A.3.2.1   *The heat equation*   For this equation, in addition to the boundary conditions, the solution must satisfy the initial conditions

$$
u(x, 0) = u_0(x), \; \forall x \in \Omega.
$$
(A.11)

As for the boundary conditions, they are essentially the same as for the boundary-value problem considered in the case of the Laplace operator. Thus, given the functions u_0, f_Ω, and g_∂ (u_0 defined in Ω, f_Ω defined in $\Omega \times [0, T]$, and g_∂ defined in $\partial\Omega \times [0, T]$), the most general initial-boundary-value problem here considered for the heat equation consists of finding a function that fulfills

$$
\frac{\partial u}{\partial t} - \Delta u = f_\Omega \text{ in } \Omega \times (0, T)
$$
$$
\alpha\frac{\partial u}{\partial n} + \beta u = g_\partial \text{ on } \partial\Omega \times (0, T)
$$
(A.12)

together with the initial conditions of Eq. (A.11).

A.3.2.2 The wave equation For this equation, in addition to the boundary conditions, the solution must satisfy the initial conditions; that is,

$$\left. \begin{array}{l} u\left(\underline{x},0\right) = u_0\left(\underline{x}\right) \\ \dfrac{\partial u}{\partial t}\left(\underline{x},0\right) = u_0'\left(\underline{x}\right) \end{array} \right\} \forall \underline{x} \in \Omega. \tag{A.13}$$

The boundary conditions in this instance are the same as for the heat equation. Thus, given the functions u_0, u_0', f_Ω, and g_∂ (u_0 and u_0' defined in Ω, f_Ω defined in $\Omega \times [0,T]$ and g_∂ defined in $\partial\Omega \times [0,T]$), the most general initial-boundary-value problem considered here for the wave equation consists of finding a function that fulfills

$$\begin{array}{rcl} \dfrac{\partial u}{\partial t} - \Delta u &=& f_\Omega \text{ in } \Omega \times (0,T) \\[2mm] \alpha\dfrac{\partial u}{\partial n} + \beta u &=& g_\partial \text{ on } \partial\Omega \times (0,T) \end{array} \tag{A.14}$$

together with the initial conditions of Eq. (A.11).

REFERENCES

1. Evans, L.C., *Partial Differential Equations, Graduate Studies in Mathematics*, Vol. 19, American Mathematical Society, Providence, RI, 1998.

2. Garabedian, P.R., *Partial Differential Equations*, Wiley, New York, 1964.

APPENDIX B

SOME RESULTS FROM THE CALCULUS

B.1 NOTATION

In what follows, Ω will be a domain of the *m-dimensional Euclidean space*. By a *partition of* Ω, we mean a collection of disjoint sub domains $\{\Omega_1, ..., \Omega_E\}$ with disjoint *closures* such that $\bar{\Omega} = \bigcup_{\varepsilon=1}^{E} \bar{\Omega}_\alpha$. Here $\bar{\Omega}$ and $\bar{\Omega}_\alpha$ stand for the *closures* of Ω and Ω_α, respectively, while the symbols $\partial\Omega$ and $\partial\Omega_\alpha$ will be used for the respective boundaries, in the mathematical sense.

When a domain Ω and one of its partitions, $\{\Omega_1, ..., \Omega_E\}$, are considered, let us write

$$\Sigma_{ij} \equiv \partial\Omega_i \cap \partial\Omega_j \text{ and } \Sigma \equiv \bigcup_{i \neq j} \Sigma_{ij}. \tag{B.1}$$

Then we refer to Σ as the *internal boundary* of Ω, and in such a case, for greater clarity, $\partial\Omega$ will be said to be its *external boundary*. We observe that the internal boundary of Ω is relative to a specified partition, whereas its external boundary is independent of the partition. It will always be assumed that the internal boundary has been oriented: that is, that a *positive* and a *negative* side have been defined on it. At

Mathematical Modeling in Science and Engineering: An Axiomatic Approach. **217**
By Ismael Herrera and George F. Pinder Copyright © 2012 John Wiley & Sons, Inc.

each point of the internal boundary, and also of the external boundary, a unit normal vector, $\underline{n} \equiv (n_1, ..., n_m)$, is defined. On $\partial\Omega$ it points toward the exterior of Ω, while on Σ it points toward the *positive side*. We observe that these concepts are well-defined everywhere on $\partial\Omega$ and Σ, except at corners and edges. That notwithstanding, we will be able to carry out the developments that follow, mainly due to the fact that the union of such corners and edges constitute a set of zero volume.

A function defined in a domain Ω, which is provided with a partition, is said to be *piecewise continuous* when its restriction to any Ω_α can be extended to $\bar{\Omega}_\alpha$ as a continuous function. When a function f defined in Ω is piecewise continuous, we define its *jump* by

$$[\![f]\!] \equiv f_+ - f_- \tag{B.2}$$

where f_+ and f_- are the limits of f, on Σ, from the positive and negative sides, respectively. We observe that $[\![f]\!]$ is well-defined on Σ, except at corners and edges.

B.2 GENERALIZED GAUSS THEOREM

We will make extensive use of a generalized version of Gauss's theorem (also known as the *Divergence theorem*). This theorem is easily derived from *Green's theorem*; furthermore, it can be interpreted as an extension to multidimensional spaces of the well-known *Fundamental Theorem of Calculus*, which applies to one-variable functions and asserts that the integral of the total derivative of a function equals the function.

Theorem 1 (Green's Theorem) *Let Ω be a domain of the m-dimensional Euclidean space, while $f(x_1, ..., x_m)$ is a function which is continuous together with its first-order partial derivatives in Ω. Then*

$$\int_\Omega \frac{\partial f}{\partial x_i}(\underline{x}) \, d\underline{x} = \int_{\partial\Omega} f(\underline{x}) \, n_i \, d\underline{x}. \tag{B.3}$$

Theorem 2 (Generalized Green's Theorem) *A straightforward generalization of this theorem is quite relevant for our purposes. Let Ω be a domain of the m-dimensional Euclidean space, while $f(x_1, ..., x_m)$ is a function which is piecewise continuous together with its first-order partial derivatives in Ω. Then*

$$\int_\Omega \frac{\partial f}{\partial x_i}(\underline{x}) \, d\underline{x} = \int_{\partial\Omega} f(\underline{x}) \, n_i \, d\underline{x} - \int_\Sigma [\![f(\underline{x})]\!] \, n_i \, d\underline{x}. \tag{B.4}$$

Proof: We write

$$\int_\Omega \frac{\partial f}{\partial x_i}(\underline{x}) \, d\underline{x} = \sum_{\alpha=1}^E \int_{\Omega_\alpha} \frac{\partial f}{\partial x_i}(\underline{x}) \, d\underline{x} \tag{B.5}$$

and then apply *Green's theorem* at each of the subdomains of the partition:

$$\int_{\Omega_\alpha} \frac{\partial f}{\partial x_i}(\underline{x}) \, d\underline{x} = \int_{\partial\Omega_\alpha} f(\underline{x}) \, n_i \, d\underline{x}. \tag{B.6}$$

Therefore,

$$\int_\Omega \frac{\partial f}{\partial x_i}(\underline{x})\, d\underline{x} = \sum_{\alpha=1}^{E} \int_{\Omega_\alpha} \frac{\partial f}{\partial x_i}(\underline{x})\, d\underline{x} = \sum_{\alpha=1}^{E} \int_{\partial\Omega_\alpha} f(\underline{x})\, n_i\, d\underline{x}. \qquad \text{(B.7)}$$

The theorem then follows from the relation:

$$\sum_{\alpha=1}^{E} \int_{\partial\Omega_\alpha} f(\underline{x})\, n_i\, d\underline{x} = \int_{\partial\Omega} f(\underline{x})\, n_i\, d\underline{x} - \int_\Sigma [\![f(\underline{x})]\!]\, n_i\, d\underline{x}. \qquad \text{(B.8)}$$

Equation (B.8) was proved as an exercise. ∎

In turn, a generalized form of Gauss's theorem is a corollary of this *generalized Green's theorem*.

Theorem 3 (Generalized Gauss Theorem) *Let Ω be a domain of the m-dimensional Euclidean space, while $\underline{u}(x_1, ..., x_m) \equiv (u_1(x_1, ..., x_m), ..., u_m(x_1, ..., x_m))$ is a vector function, which is piecewise continuous together with its first-order partial derivatives in Ω. Then*

$$\int_\Omega \nabla \cdot \underline{u}\, dx = \int_{\partial\Omega} \underline{u} \cdot \underline{n}\, dx - \int_\Sigma [\![\underline{u}]\!] \cdot \underline{n}\, dx. \qquad \text{(B.9)}$$

Proof: Using Green's generalized theorem;

$$\begin{aligned}
\int_\Omega \nabla \cdot \underline{u}\, d\underline{x} &= \sum_{i=1}^{m} \int_\Omega \frac{\partial u_i}{\partial x_i}(\underline{x})\, d\underline{x} \\
&= \sum_{i=1}^{m} \left\{ \int_{\partial\Omega} u_i n_i\, d\underline{x} - \int_\Sigma [\![u_i]\!]\, n_i\, d\underline{x} \right\} \\
&= \int_{\partial\Omega} \underline{u} \cdot \underline{n}\, d\underline{x} - \int_\Sigma [\![\underline{u}]\!] \cdot \underline{n}\, d\underline{x} \qquad \text{(B.10)}
\end{aligned}$$

We observe that in particular when the vector field $\underline{u}(x_1, ..., x_m)$ is continuous in Ω, this theorem adopts the standard form of Gauss's theorem. Indeed, when $[\![\underline{u}]\!] \cdot \underline{n} = 0$ on Σ, Eq. (B.9) reduces to

$$\int_\Omega \nabla \cdot \underline{u}\, dx = \int_{\partial\Omega} \underline{u} \cdot \underline{n}\, d\underline{x}. \qquad \text{(B.11)}$$

∎

APPENDIX C

PROOF OF THEOREM

The purpose of this appendix is to prove the following theorem.

Theorem 1 *Let the function $\psi(\underline{x}, t)$ be piecewise-continuous together with its first-order partial derivatives in a suitable space-time domain. When $B(t)$ is the domain occupied by a body at any time t, we define*

$$E(t) \equiv \int_{B(t)} \psi(\underline{x}, t)\, d\underline{x}. \tag{C.1}$$

Then

$$\frac{dE}{dt}(t) = \int_{B(t)} \left\{ \frac{\partial \psi}{\partial t} + \nabla \cdot (\psi \underline{v}) \right\} d\underline{x} + \int_{\Sigma(t)} [\![\psi(\underline{v} - \underline{v}_\Sigma)]\!] \cdot \underline{n}\, d\underline{x} \tag{C.2}$$

Proof: To understand the proof that follows, we first recall a well-known result from calculus, which permits computing the derivative of the integral of an intensive property when the domain $\Omega(t)$ of integration changes with time, in such a way that its boundary $\partial\Omega(t)$ moves with a certain velocity, to be denoted by $\underline{v}_\partial$ relationship.

Mathematical Modeling in Science and Engineering: An Axiomatic Approach.
By Ismael Herrera and George F. Pinder Copyright © 2012 John Wiley & Sons, Inc.

When $\psi\,(\underline{x},t)$ is continuous together with its first-order partial derivatives and

$$F\,(t) \equiv \int_{\Omega(t)} \psi\,(\underline{x},t)\,d\underline{x} \tag{C.3}$$

then,

$$\frac{dF}{dt}\,(t) = \int_{\Omega(t)} \frac{\partial \psi}{\partial t}\,d\underline{x} + \int_{\partial\Omega(t)} \psi \underline{v}_\partial \cdot \underline{n}\,d\underline{x} \tag{C.4}$$

Some remarks should be made. Generally, the *particle* and *shock velocities*, $\underline{v}$ and $\underline{v}_\Sigma$, may be different. This implies that in such a case, some particles of a body may go across the *shock*. Due to this fact, parts of a subbody that at a given time lies entirely in one side of a shock may go across the shock and, later, lie at the other side of the shock. That must be taken into account when making the balance of *extensive properties*. Given the time-dependent domain $B\,(t)$, we take a time-dependent partition $\{\Omega_1\,(t)\,,...,\Omega_E\,(t)\}$ satisfying at each time the following properties:

- For each α, $\psi\,(\underline{x},t)$ is continuous, together with its first-order partial derivatives, on $\Omega_\alpha\,(t)$. Because of these continuity assumptions, for each α, $\partial\Omega_\alpha\,(t)$ cannot cross the *shock*, $\Sigma\,(t)$; and

- Writing $\partial\Omega_\alpha$ is the union of two disjoint sets: $\Sigma\cap\partial\Omega_\alpha$ and $\partial\Omega_\alpha - (\Sigma\cap\partial\Omega_\alpha)$. The set $\partial\Omega_\alpha - (\Sigma\cap\partial\Omega_\alpha)$ is taken so that it moves with the velocity of the particles, $\underline{v}$. As for the set $\Sigma\cap\partial\Omega_\alpha$, it necessarily moves with the velocity of the shock, $\underline{v}_\Sigma$.

Then we write

$$E\,(t) = \sum_{\alpha=1}^{E} \int_{\Omega_\alpha(t)} \psi\,(\underline{x},t)\,d\underline{x}. \tag{C.5}$$

Furthermore, applying Eq. (C.4), we have

$$\frac{d}{dt}\int_{\Omega_\alpha(t)} \psi\,d\underline{x} = \int_{\Omega_\alpha(t)} \frac{\partial\psi}{\partial t}\,d\underline{x} + \int_{\partial\Omega_\alpha - (\Sigma\cap\partial\Omega_\alpha)} \psi\underline{v}\cdot\underline{n}\,d\underline{x} + \int_{\Sigma\cap\partial\Omega_\alpha} \psi\underline{v}_\Sigma\cdot\underline{n}\,d\underline{x}. \tag{C.6}$$

Now

$$\int_{\partial\Omega_\alpha - (\Sigma\cap\partial\Omega_\alpha)} \psi\underline{v}\cdot\underline{n}\,d\underline{x} = \int_{\partial\Omega_\alpha} \psi\underline{v}\cdot\underline{n}\,d\underline{x} - \int_{\Sigma\cap\partial\Omega_\alpha} \psi\underline{v}\cdot\underline{n}\,d\underline{x}$$

$$= \int_{\Omega_\alpha} \nabla\cdot(\psi\underline{v})\,d\underline{x} - \int_{\Sigma\cap\partial\Omega_\alpha} \psi\underline{v}\cdot\underline{n}\,d\underline{x} \tag{C.7}$$

Therefore,

$$\frac{d}{dt}\int_{\Omega_\alpha(t)} \psi\,d\underline{x} = \int_{\Omega_\alpha(t)} \left\{\frac{\partial\psi}{\partial t} + \nabla\cdot(\psi\underline{v})\right\}d\underline{x} + \int_{\Sigma\cap\partial\Omega_\alpha} \psi\,(\underline{v}_\Sigma - \underline{v})\cdot\underline{n}\,d\underline{x} \tag{C.8}$$

and

$$\frac{dE}{dt}(t) = \int_{B(t)} \left\{ \frac{\partial \psi}{\partial t} + \nabla \cdot (\psi \underline{v}) \right\} d\underline{x} + \sum_{\alpha=1}^{E} \int_{\Sigma \cap \partial \Omega_\alpha} \psi \left(\underline{v}_\Sigma - \underline{v} \right) \cdot \underline{n} \, d\underline{x} \quad \text{(C.9)}$$

.This equation reduces to Eq. (C.2) when use is made of the fact that

$$\sum_{\alpha=1}^{E} \int_{\Sigma \frown \partial \Omega_\alpha} \psi \left(\underline{v}_\Sigma - \underline{v} \right) \cdot \underline{n} \, d\underline{x} = \int_{\Sigma(t)} [[\psi \left(\underline{v} \right) - \underline{v}_\Sigma]] \cdot d\underline{x}. \quad \text{(C.10)}$$

∎

APPENDIX D

THE BOUNDARY LAYER
INCOMPRESSIBILITY APPROXIMATION

Consider the boundary-value problem in which we seek a function $h(\underline{x}, t)$ that satisfies

$$S_S \frac{\partial h}{\partial t} - K\Delta h = 0, \quad \forall \underline{x} \in \Omega, \quad \text{and} \quad t > 0 \tag{D.1}$$

subject to the boundary conditions

$$h(\underline{x}, t) = h_\partial(\underline{x}), \quad \forall \underline{x} \in \Omega, \quad \text{and} \quad t > 0 \tag{D.2}$$

and initial conditions

$$h(\underline{x}, 0) = h_0(\underline{x}), \quad \forall \underline{x} \in \Omega. \tag{D.3}$$

Here, both h_∂ and h_0 are prescribed functions.

On the other hand, let the *incompressible solution* be such that it fulfills

$$K\Delta \bar{h} = 0, \quad \forall \underline{x} \in \Omega, \quad \text{and} \quad t > 0 \tag{D.4}$$

together with the boundary conditions

Mathematical Modeling in Science and Engineering: An Axiomatic Approach.
By Ismael Herrera and George F. Pinder Copyright © 2012 John Wiley & Sons, Inc.

Then the difference between these two functions, $\delta\left(\underline{x},t\right) \equiv h\left(\underline{x},t\right) - \bar{h}\left(\underline{x}\right)$, satisfies

$$S_S \frac{\partial \delta}{\partial t} - K \Delta \delta = 0, \quad \forall \underline{x} \in \Omega, \text{ and } t > 0 \tag{D.6}$$

subjected to the boundary conditions

$$\delta\left(\underline{x},t\right) = 0, \quad \forall \underline{x} \in \Omega, \text{ and } t > 0 \tag{D.7}$$

and initial conditions

$$\delta\left(\underline{x},0\right) = h_0\left(\underline{x}\right) - \bar{h}\left(\underline{x}\right), \quad \forall \underline{x} \in \Omega. \tag{D.8}$$

Therefore,

$$\delta\left(\underline{x},t\right) = \sum_{n=0}^{\infty} A_n W^n\left(\underline{x}\right) e^{-\frac{K}{S_S}\lambda_n t}, \quad \forall \underline{x} \in \Omega \text{ and } t > 0. \tag{D.9}$$

Here W^n and λ_n, $n = 0, 1, ...$, are the proper functions (eigenfunctions) and proper values (eigenvalues) of the problem

$$-\Delta W = \lambda_n W \text{ in } \Omega \tag{D.10}$$

subjected to the boundary conditions

$$W\left(\underline{x}\right) = 0 \text{ for } \forall \underline{x} \in \partial\Omega. \tag{D.11}$$

This problem has the property that all the proper values λ_n, $n = 0, 1, ...$, are positive. We order them so that

$$0 < \lambda_0 < \lambda_1 < \cdots < \lambda_n < \cdots. \tag{D.12}$$

If we normalize the proper functions, and δ_{nm} is the Kronecker delta, then

$$\int_{\Omega} W^n W^m dx = \delta_{nm}. \tag{D.13}$$

In this case,

$$A_n = \int_{\Omega} \left(h_0 - \bar{h}\right) W^n\, dx \tag{D.14}$$

and

$$\left\| h_0 - \bar{h}\right\| = \sqrt{\sum_{n=0}^{\infty} A_n^2} \tag{D.15}$$

where

$$\left\| h_0 - \bar{h}\right\| \leq \sqrt{\int_{\Omega} \left(h_0 - \bar{h}\right)^2 dx}. \tag{D.16}$$

Then it can be shown that

$$\|\delta\left(\underline{x},t\right)\|^2 = \sum_{n=0}^{\infty} A_n^2 e^{-\frac{2K}{S_S}\lambda_n t} \le e^{-\frac{2K}{S_S}\lambda_0 t}\sum_{n=0}^{\infty} A_n^2 = \left\|h_0 - \bar{h}\right\|^2 e^{-\frac{2K}{S_S}\lambda_0 t}. \quad \text{(D.17)}$$

Therefore,

$$\|\delta\left(\underline{x},t\right)\| \le \left\|h_0 - \bar{h}\right\| e^{-\frac{K}{S_S}\lambda_0 t}. \quad \text{(D.18)}$$

Clearly, the norm of the error, $\|\delta\left(\underline{x},t\right)\|$, goes to zero when $S_S \to 0$. Furthermore, we could define the thickness, ζ, of the boundary layer as the time required for the norm of the error to be less than a certain factor, for example, 0.1, in which case

$$\zeta \le \frac{S_S}{K\lambda_0}\log\ 10. \quad \text{(D.19)}$$

For example, if Ω is a cube whose sides equal L, then

$$\lambda_0 = \frac{3L^2}{\pi^2}. \quad \text{(D.20)}$$

APPENDIX E

INDICIAL NOTATION

E.1 GENERAL

In elementary calculus and analytic geometry, it is customary to write x, y, and z for the coordinates of three-dimensional *Euclidean space*. However, there are many advantages to using subscripts to distinguish such coordinates, in which case we write x_1, x_2, and x_3 instead (sometimes superscripts are used in place of subscripts); in summary, the notation x_i, with $i = 1, 2, 3$, represents the coordinates of a point. This defines number systems, (x_1, x_2, x_3), constituted by three elements. More generally, we may consider number systems represented by any number (n) of real numbers that will be denoted by x_i, with $i = 1, ..., n$. The total set of such number systems is referred to as the *n-dimensional* Euclidean space and is denoted by R^n. Number systems such as those considered prior to Section 2.6, which depend on only one index, are said to be *first-order systems*. We also consider *second-order systems*, such as (a_{ij}), with $i, j = 1, ..., n$, in which the numbers of the system depend on two indices. Given any first-order system x_i, with $i = 1, ..., n$, we can associate with it the vector $\underline{x} \equiv (x_1, ..., x_n)$; this vector may be interpreted as the *position vector* of a point in a rectangular *Cartesian system* of coordinates. Thus, to each point in an

Mathematical Modeling in Science and Engineering: An Axiomatic Approach.
By Ismael Herrera and George F. Pinder Copyright © 2012 John Wiley & Sons, Inc.

dimensional Euclidean space there corresponds one and only one *n-dimensional* vector: $\underline{x} \equiv (x_1, ..., x_n)$. In particular, R^3 is the three-dimensional physical space in which the points are given by the system numbers x_i, with $i = 1, 2, 3$.

In our notation, vectors are underlined; thus, we write $\underline{x}$ for a vector and understand that $\underline{x} \equiv (x_1, ..., x_n)$ when the vector $\underline{x}$ is *n-dimensional*. The *scalar product* (or *inner product*) of the vectors $\underline{x}$ and $\underline{y} \equiv (y_1, ..., y_n)$ is defined by

$$\underline{x} \cdot \underline{y} \equiv x_1 y_1 + \cdots + x_n y_n = \sum_{i=1}^{n} x_i y_i. \tag{E.1}$$

This is not the only manner in which the scalar product of two vectors can be defined; thus, to be unambiguous the scalar product defined by Eq. (E.1) is sometimes referred to as the *Euclidean inner product*.

The so-called \summation convention" consists of omitting the sum sign Σ in expressions such as Eq. (E.1); more precisely, we agree that a Σ sign is tacitly understood whenever an index is repeated in a single term; the sum is carried out over the range of the repeated index. Furthermore, such an index is said to be *contracted*; it should be noted that the letter used to denote a contracted index is immaterial. Indeed, for example, in this notation the inner product of the vectors, $\underline{x}$ and $\underline{y}$ above is simply

$$x_i y_i = x_j y_j = x_m y_m. \tag{E.2}$$

When

$$\underline{a} \equiv (a_{ij}), \quad i, j = 1, ..., n \tag{E.3}$$

is a *second-order system* of numbers, we have

$$a_{ij} x_i y_j \equiv \sum_{i,j=1}^{n} a_{ij} x_i y_j. \tag{E.4}$$

It should be stressed that the writing of many expressions is very greatly simplified when the *indicial notation* is combined with the summation convention; indeed, for example, when $n = 3$, we have

$$a_{ij} x_i y_j = \begin{array}{l} a_{11} x_1 y_1 + a_{12} x_1 y_2 + a_{13} x_1 y_3 \\ +a_{21} x_2 y_1 + a_{22} x_2 y_2 + a_{23} x_2 y_3 \\ +a_{31} x_3 y_1 + a_{32} x_3 y_2 + a_{33} x_3 y_3. \end{array} \tag{E.5}$$

The summation convention is used throughout, except where explicitly stated otherwise. Also, when considering second order systems it is assumed that, generally, $a_{ij} \neq a_{ji}$.

E.2   MATRIX ALGEBRA

There are two ways of associating a vector to a *first order system of numbers*. Indeed, let x_i, with $i = 1, ..., n$, be such a system; then one can associate with it the *row vector:* $\underline{x} \equiv (x_1, ..., x_n)$, as we did before; or the *column vector,* given by the transpose of such a vector: that is,

$$
\underline{x}^T \equiv \begin{pmatrix} x_1 \\ \cdot \\ \cdot \\ \cdot \\ x_n \end{pmatrix}. \tag{E.6}
$$

Similarly, when we associate a matrix to a second-order system such as that given by Eq. (E.3), the associated matrices $\underline{\underline{a}}$ may be

$$
\underline{\underline{a}} \equiv \begin{pmatrix} a_{11} & \cdot & \cdot & \cdot & a_{1n} \\ & \cdot & \cdot & \cdot & \cdot \\ \cdot & \cdot & \cdot & \cdot & \cdot \\ \cdot & \cdot & \cdot & \cdot & \cdot \\ a_{n1} & \cdot & \cdot & \cdot & a_{nn} \end{pmatrix} \tag{E.7}
$$

or

$$
\underline{\underline{a}}^T \equiv \begin{pmatrix} a_{11} & \cdot & \cdot & \cdot & a_{n1} \\ & \cdot & \cdot & \cdot & \cdot \\ \cdot & \cdot & \cdot & \cdot & \cdot \\ \cdot & \cdot & \cdot & \cdot & \cdot \\ a_{1n} & \cdot & \cdot & \cdot & a_{nn} \end{pmatrix}. \tag{E.8}
$$

To avoid this ambiguity, given a matrix, the standard rule applied when associating with it a second-order system consists of choosing the *first index in the number system equal to the row position in the matrix.* Then the correspondence between matrices and second-order systems is one-to-one. In particular, the equation

$$
\underline{\underline{a}}\underline{x} = \underline{y} \tag{E.9}
$$

is equivalent to

$$
a_{ij}x_j = y_i, \quad i = 1, ..., n. \tag{E.10}
$$

It should be noted that the indicial notation is more explicit and simpler than the matrix notation.

Clearly, a matrix $\underline{\underline{a}}$ is symmetric if, and only if,

$$
a_{ij} = a_{ji}, \quad \forall i, j = 1, ..., n. \tag{E.11}
$$

Moreover, it is skew-symmetric if, and only if,

$$
a_{ij} = -a_{ji}, \quad \forall i, j = 1, ..., n. \tag{E.12}
$$

A matrix that plays a very special role in linear algebra is the *identity matrix*, denoted by $\underset{=}{I}$. We notice that

$$\underset{=}{I} = (\delta_{ij}) . \tag{E.13}$$

Here δ_{ij} is the *Kronecker delta*, which is defined by

$$\delta_{ij} = \begin{cases} 1 \text{ when } i = j \\ 0 \text{ when } i \neq j. \end{cases} \tag{E.14}$$

E.3 APPLICATIONS TO DIFFERENTIAL CALCULUS

First, we notice that if $u\,(x_1, ..., x_n)$ is a function of n independent variables, it can be thought of as a function of the position vector $\underline{x} \equiv (x_1, ..., x_n)$ in the n-*dimensional* Euclidean space, R^n.

In differential calculus the symbols $\nabla\,(\cdot)$ and $\Delta\,(\cdot)$ are frequently used to denote certain differential operators. When they are expressed by means of indicial notation, their meaning is very precise and their algebra is easy to carry out. The definition of the Laplacian operator $\Delta\,(\cdot)$ is straightforward when it is given using indicial notation. When, furthermore, the summation convention is also applied, it can be written as $\partial^2/\partial x_i \partial x_i$. The action of this operator on a function of n independent variables, $u\,(x_1, ..., x_n)$, is given by the identity

$$\Delta u = \frac{\partial^2 u}{\partial x_i\,\partial x_i} = \sum_{i=1}^{n} \frac{\partial^2 u}{\partial x_i\,\partial x_i} = \frac{\partial^2 u}{\partial x_1\,\partial x_1} + \cdots + \frac{\partial^2 u}{\partial x_n\,\partial x_n}. \tag{E.15}$$

The operator $\nabla\,(\cdot)$ may be thought of as a vector-valued differential operator, and with the help of the indicial notation, it can be written as

$$\nabla\,(\cdot) \equiv \left(\frac{\partial\,(\cdot)}{\partial x_1}, ..., \frac{\partial\,(\cdot)}{\partial x_n} \right). \tag{E.16}$$

When it is applied to a real-valued function of n independent variables, $u\,(x_1, ..., x_n)$, it yields the vector-valued function

$$\nabla u \equiv \left(\frac{\partial u}{\partial x_1}, ..., \frac{\partial u}{\partial x_n} \right). \tag{E.17}$$

We observe that this vector is the *gradient* of the function $u\,(x_1, ..., x_n)$; it has the property that its inner product with a unit vector yields the directional derivative in the direction of such a vector. Thus, the directional derivative in the direction of a unit vector $\underline{n}$ is given by

$$\frac{\partial u}{\partial n} = \frac{\partial u}{\partial x_i} n_i = (\nabla u) \cdot \underline{n}. \tag{E.18}$$

In particular, when $\underline{n}$ is the unit normal vector to a surface, $\frac{\partial u}{\partial n}$ is usually called the *normal derivative*.

If $\underline{u}(x_1, ..., x_n) = \underline{u}(\underline{x})$ is a vector-valued function, the operator ∇ can be applied to it in several ways. First, $\nabla \underline{u}$ is defined to be a matrix associated the *second-order system* $(\partial u_i / \partial x_j)$, with $i, j = 1, ..., n$, which itself is a function of position. Thus, we write

$$\nabla \underline{u} \equiv \frac{\partial u_i}{\partial x_j}, \quad i, j = 1, ..., n. \tag{E.19}$$

However, this definition is ambiguous since the indexes i and j do not have an order in Eq. (E.19). Therefore, it is necessary to choose an order for them, in order to define $\nabla \underline{u}$ as a matrix. The standard definition of $\nabla \underline{u}$ is such that the following expression, frequently used for evaluating the normal derivative, holds:

$$\frac{\partial \underline{u}}{\partial n} = (\nabla \underline{u}) \, \underline{n}. \tag{E.20}$$

That is, when the matrix $\nabla \underline{u}$ is applied to the unit normal vector, the normal derivative of the vector-valued function $\underline{u}$ is obtained. When this is the case, Eqs. (E.9), (E.10), (E.18), and (E.20), together, imply that in Eq. (E.19) i is the first index and j is the second. Hence, the standard definition is

$$\nabla \underline{u} \equiv \underline{\underline{a}} \equiv (a_{ij}) \quad \text{with } a_{ij} \equiv \frac{\partial u_i}{\partial x_j}. \tag{E.21}$$

Once this has been established, the matrix $\underline{\underline{a}} \equiv \nabla \underline{u}$ can be constructed in the usual manner: The first index corresponds to the number of the matrix row, while the second defines the column.

A second way in which the differential operator $\nabla (\cdot)$ can be applied to a vector-valued function, $\underline{u}$, mimics the manner in which the inner product between two vectors is evaluated, albeit in this case the inner product is between the differential operator $\nabla (\cdot)$ and the vector $\underline{u}$. It is given by

$$\nabla \cdot \underline{u} \equiv \frac{\partial}{\partial x_i} u_i = \frac{\partial u_i}{\partial x_i} = \sum_{i=}^{n} \frac{\partial u_i}{\partial x_i}. \tag{E.22}$$

Thus, $\nabla \cdot \underline{u}$ is the *divergence* of the vector-valued function, $\underline{u}$.

We observe that in many cases a clear advantage of the indicial notation is that it is more precise and easier to apply. For example, as explained before, the expression $\nabla \cdot (\psi \nabla \underline{u})$ is ambiguous. However, $\nabla \underline{u}$ has been defined as the matrix of Eq. (E.21), and therefore the precise meaning of $\nabla \cdot (\psi \nabla \underline{u})$ in indicial notation is

$$\nabla \cdot (\psi \nabla \underline{u}) \equiv \frac{\partial}{\partial x_j} \left(\psi \frac{\partial u_i}{\partial x_j} \right). \tag{E.23}$$

Then, using elementary formulas for the derivatives of a product, one obtains

$$\frac{\partial}{\partial x_j} \left(\psi \frac{\partial u_i}{\partial x_j} \right) = \psi \frac{\partial^2 u_i}{\partial x_j \, \partial x_j} + \frac{\partial u_i}{\partial x_j} \frac{\partial \psi}{\partial x_j}. \tag{E.24}$$

When the differential operators $\nabla \left(\cdot \right)$ and $\Delta \left(\cdot \right)$ are introduced, this can be written as

$$\nabla \cdot (\psi \nabla \underline{u}) = \psi \Delta \underline{u} + \nabla \underline{u} \nabla \psi. \tag{E.25}$$

We observe that $\nabla \underline{u}$ is a matrix while $\nabla \psi$ is a vector; therefore, the product $\nabla \underline{u} \nabla \psi$ is well-defined. On the other hand, the expression $\nabla \psi \nabla \underline{u}$ does not make sense. Hence, when the notation based on the symbols ∇ and Δ was introduced, we were led to algebraic operations that are sensitive to order, and their handling requires extreme care. On the other hand, when, instead, the indicial notation is used this does not happen. Indeed, the operation $\nabla \underline{u} \nabla \psi$, in indicial notation, corresponds to

$$\frac{\partial u_i}{\partial x_j} \frac{\partial \psi}{\partial x_j} = \frac{\partial \psi}{\partial x_j} \frac{\partial u_i}{\partial x_j}, \quad i = 1, ..., n. \tag{E.26}$$

However, in the latter expressions the product is commutative.

In conclusion, when indicial notation was used we were able to apply in our derivations elementary differential formulas and ordinary products that are defined between numbers, something that was not possible when the differential operators $\nabla \left(\cdot \right)$ and $\Delta \left(\cdot \right)$ were introduced. However, before finishing this appendix, it should be said that both notations have their own merits and are used extensively, but when carrying out algebraic developments, the indicial notation has clear advantages and should be preferred. A useful exercise consists of translating formulas from one notation to the other.

INDEX

Mathematical Modeling in Science and Engineering: An Axiomatic Approach.
By Ismael Herrera and George F. Pinder Copyright © 2012 John Wiley & Sons, Inc.